AF880932

Building our Digital World:

DELIVERING SOFTWARE AND SYSTEMS IN SUPPORT OF OUR DIGITAL ECONOMY

To my wonderful wife, Carolyn, and to our family, who have all demonstrated remarkable patience and sacrifice throughout the lengthy process and countless hours required to bring the Building Our Digital World series to life. I love you all.

Building Our Digital World

SDLC Foundations

An introductory and historical guide
to software and systems development

BY CECIL GARY RUP

Cover Artwork

Along the way to publishing this series of books, I discovered two important things. One, it takes a good team to develop a quality product. Second, even seemingly trivial tasks, like designing the book covers, can turn into large projects in their own right. Such was the case with designing the Building our Digital World Series book covers, and the artwork titled *Engaged in Our Digital World.*

The design concept began with the notion that we are all connected to a digital world; a world that is built by the efforts of software and IT-based systems developers. My book designer, Adam Hay, of Adam Hay Studio, and I saw several artificial intelligence (AI) based artworks that suggested such connections. While I liked the imagery of some of the AI-based artworks, they typically included pictures of androids or robots. I wanted to make the point that it's humans who are increasingly connected to our digital world. Besides the digital connections, I also wanted to convey the role software programming plays in forming such digital connections.

We could not find an existing commercial illustration and graphic that evoked the image I was seeking. So, I turned to the folks at Unzipped Designs, Victoria Moore and James Neely, to see if they could come up with a custom design for us. Victoria and James both work as artists in 3d modeling formats. James is a 3D creature and character artist who holds two patents for digital sculpting in product designs and the use of 3D models for pre-visualization of product packaging concepts for marketing purposes. Victoria is a software programmer who also develops 3d avatars for gamers. A perfect fit!

The final artwork displays a couple walking together with a digital rain immersing them in the virtualized digital world. The two individuals are avatars, who represent real-world humans in their digital interactions within this 'other' world. The programming code, lying upon the water-soaked ground, provides the foundation for their digital interactions.

Are the human-directed avatars and their digital world real, or imagined? I submit it's increasingly difficult to tell. After all, who as a gamer doesn't believe that the avatars they use, and those that surround them in their digital worlds, at least for the time of their play, represent reality and a real place where they reside and interact with others? I hope you enjoy the artwork as much as we enjoyed collaborating on the design.

Copyright ©2020 by Cecil Gary Rupp, Chatfield Publishing Company.

All rights reserved. The Author retains intellectual, production, and distribution rights to all versions, derivatives, and revisions of this book, whether now in existence or to be created in the future, including all copyrights in all languages, in all known and unknown forms, media, or means of expression, all rights to display, perform, reproduce, modify, merchandise, trademark, or otherwise commercially exploit this book and the Building Our Digital World series of books.

No part of this book may be reproduced or transmitted in any form or by any means, electronic or mechanical, including photocopying, recording, or any information storage and retrieval system, without permission in writing from the Author and his publisher, except in the case of brief quotations embedded in industry or media articles and reviews.

Every effort has been made by the author in the preparation of this book to ensure the accuracy of the information and citations presented. Nevertheless, the information contained in this book is sold without warranty, either express or implied, and the author assumes no responsibility for errors or omissions. Neither the author or the publisher, or its dealers and distributors, will be held liable for any damages caused or alleged to have been caused directly or indirectly by this book.

The Author has endeavored to provide trademark and registration information about all the companies and products mentioned in this book by the appropriate use of trademark and registration marks. However, the Author cannot guarantee the accuracy of this information.

Draft Copy Editor: Linda Orlando
Book Design: Adam Hay of Adam Hay Studio
Cover Artwork: Victoria Moore and James Neely of Unzipped Designs
Proofreader: Joe McDaniel of Book Crafters
Final Production Editor: Alexandra O'Connell of Your Resident Wordsmith
Marketing Specialist: Mary Walewski of Buy The Book Marketing
Website Design: Victoria Moore of Unzipped Designs

ISBN: 978-1-7341478-0-3

First Edition: April 2020

1 2 3 4 5 6 7 8 9 10

Contents

Building our Digital World - Series Introduction 22

SCOPE OF THE SDLC SERIES 24

TARGETED READERSHIP 24

MAKING THE CASE FOR SDLC FRAMEWORKS AND PROCESSES 25

A TIME OF FLUX IN OUR INDUSTRY 26

IT DIDN'T HAVE TO END THAT WAY! 27

01. Key SDLC Concepts 28

CHAPTER OBJECTIVES 28

AUTHORS INTRODUCTION 29

WHAT IS AN SDLC? 31

AGILE PROCESSES AUGMENT SDLC PROCESSES 31

WHAT IS AN SDLC PROCESS 32

Basic SDLC Concepts versus Commercialized Approaches 33

SDLC Preferences 34

IT Governance 34

SDLC as a Framework 34

SDLCS AS KNOWLEDGE ASSETS AND INTELLECTUAL PROPERTY 35

TRADITIONAL SDLC PHASES 35

THE AGILE SDLC MODEL 36

TRADITIONAL SDLC WATERFALL MODELS VERSUS AGILE PRACTICES 37

Contents

AGILE AS AN SDLC IS NOT ALL-INCLUSIVE **37**

What's in an SDLC Process? 38

Protecting System Code, Documentation, and Other Artifacts 38

PROJECT PLANNING AND MANAGEMENT **39**

Project Integration and Tailoring 39

Project Management in an Agile World 40

LIFE CYCLE METHODS ARE FOR MANAGING CHANGE **41**

Enterprise Change Model (ECM) 41

Corporate Strategy Drives Change, Realized via Programs and Projects 41

An Enterprise View to Change 42

EXPANDED VIEW OF A SYSTEM **42**

An Enterprise Life cycle Change Management Problem 43

MULTIPLE SDLC PATHS **43**

Integrating PMLC and SDLC Processes 43

SDLCS FOR COMPETITIVE ADVANTAGE **44**

SDLC PROCESS IMPROVEMENTS AND MATURITY **44**

REUSABLE SDLC TEMPLATES **45**

RIGHT-SIZE YOUR SDLC IMPLEMENTATION **45**

CHAPTER SUMMARY **46**

02. Evolving Life Cycle Development Approaches 47

CHAPTER OBJECTIVES **47**

AUTHOR'S INTRODUCTION **48**

SDLC AS AN ACRONYM **49**

THE EVOLUTION OF SDLCS **50**

Applications and Technologies Drove the Evolution 50

Strengths of the Traditional Development Model 50

Limitations with Traditional Development Methods 51

Modern Software Development Methods and Tools 52

Incremental, Iterative, and Prototyping Approaches 53

Software Development Focus Shifts to Systems Development 53

Using Frameworks to Support SDLC Frameworks for Agility 53

FOUNDATIONAL SDLC PROCESSES **54**

TRADITIONAL/ WATERFALL SDLC MODEL **55**

Traditional Waterfall Model—Covering the Basics 55

Waterfall is not Dead—Yet 55

Waterfall Model Process Overview 56

Industry Standards governing the Traditional Model 57

Traditional Waterfall Phases 57

Benefits of Traditional Waterfall Concepts 59

Waterfall Model—Uses 59

Waterfall Model—Scope Creep 59

Waterfall Model—Defining the Requirements 60

Waterfall Model—Solution Applications 60

Top-down, Waterfall Model—Pros and Cons 60

Waterfall Model—Challenges 60

Waterfall Model—Summary 61

PROTOTYPING MODEL **62**

Covering the Basics 62

Process Overview 63

Usability Engineering 64

Types of Prototypes 64

Solution Applications 65

Pros and Cons of Prototyping 65

Contents

Summary 65

IID MODEL 66

Two Development Strategies in One 66

Covering the Basics of IID 66

IID Unified Process 67

IID Release Approaches 68

Process Overview 69

IID is built on PDCA/PDSA Concepts 69

Timeboxed Iterations 70

The Importance of Customer Feedback 70

Evolving Applications 71

Pros and Cons of IID 71

IID Section Summary 72

RAD MODEL 72

Covering the Basics 72

Supports BPR and Continuous Process Improvements 72

A Tool-based Agile Strategy 73

Process Overview 73

Solution Applications 75

Pros and Cons of RAD 75

RAD Section Summary 75

BIG BANG MODEL 76

Covering the Basics 76

Process Overview 76

Solution Applications 77

Pros and Cons of the Big Bang Model 77

Big Bang Section Summary 78

AGILE MODELS 78

Introduction 87

Covering the Basics 87

Process Overview 87

MANIFESTO FOR AGILE SOFTWARE DEVELOPMENT 78

PRINCIPLES BEHIND THE AGILE MANIFESTO 78

Solution Applications 80

Pros and Cons of Agile SDLC Methods 80

Modern Agile 80

Agile Section Summary 83

CHAPTER SUMMARY 83

03. SDLC Process Improvement Approaches 85

CHAPTER OBJECTIVES 85

AUTHOR INTRODUCTIONS 86

AGILE-FRIENDLY PRACTICES 86

Test and Behavior Driven Development 86

DevOps (Development-Operations) 87

Theory of Constraints 88

Continuous Integration (CI) 89

Continuous Testing (CT) 90

Continuous Delivery & Deployment (CD/D) 91

Recap of DevOps and SDLC Automation Concepts 92

ADOPTED LEAN DEVELOPMENT STRATEGIES 93

Lean *93*

Just in Time (JIT) *94*

KANBAN *94*

Kaizen 94

Lean Software Development 95

Scrumban 95

Model-Driven Development (MDD) 95

Feature-Driven Development (FDD) 96

Six Sigma 96

Six Sigma Applications in Software Development 97

CHAPTER SUMMARY **98**

04. Representative SDLC Methodologies 99

CHAPTER OBJECTIVES **99**

AUTHOR'S INTRODUCTION **100**

Diverse SDLC Approaches 102

Software Development and Documentation through Models 100

Agile Unified Process (AUP) 101

Cynefin Framework 101

Disciplined Agile (DA) 102

Dynamic Systems Development Model (DSDM) 104

eXtreme Programming (XP) 104

Open Unified Process (OpenUP) 106

Outside–In Software Development 106

CMM 107

CMMI 108

PSP and TSP 109

RUP 109

SAFe® 109

Scrum Agile Methodology 110

Spiral SDLC Model 111

V-Shaped SDLC Model 112

CHAPTER SUMMARY **114**

05. Making a Case for the Hybrid SDLC Model **115**

CHAPTER OBJECTIVES **115**

AUTHOR'S INTRODUCTION **116**

TRADITIONAL SDLC PRACTICES **117**

Not your Grandfather's Waterfall 118

Waterfall Feels Natural 118

Best Fit 118

Documentation and Configuration Management 119

How things can go wrong with waterfall 119

MANAGING PERCEPTIONS **120**

MANAGING SCOPE CREEP **120**

Manage Change *121*

AGILE **122**

Prototype Development 122

Prototypes are a throwaway effort *122*

IID 123

Trial and Error *123*

Architecture and Design Enhancements *123*

RAD 123

RAD Tools *123*

Emphasis on Speed *124*

HOW THINGS CAN GO WRONG WITH AGILE **124**

HYBRID SDLC MODELS **125**

A View of Sustainability 125

DESIGN, BUILD & TEST FOR PERFORMANCE **126**

CHAPTER SUMMARY **126**

06. Turning Strategy into Action **128**

CHAPTER OBJECTIVES **128**

AUTHOR'S INTRODUCTION **129**

WHY CHANGE **130**

Business Drivers 130

Business Strategy *131*

Strategic Planning 131

Strategic Goals and Objectives 132

Course Corrections 132

Metrics—what "Good" Looks Like 132

Change is Inevitable 132

STRATEGIC PLANNING DRIVES PORTFOLIO, PROGRAM, AND PROJECT PLANNING **133**

Portfolio Management 134

Program and Project Management 134

Where Organizations Get It Wrong *135*

How to Get It Right *145*

MANAGING RISKS **137**

Risks with External Bids 137

Risks with Initial Product Concept Assessments 137

Risks with Contingency Plans 137

Risks with Deferring Development Costs 138

PROJECT FUNDING AND RESOURCE PRIORITIES **138**

OUTSOURCING CUSTOM DEVELOPMENT WORK **138**

COTS-BASED SOLUTIONS 139

FIRM FIXED-PRICE CONTRACTS 139

Dangers in Shortchanging the Life Cycle Process 140

CHAPTER SUMMARY 141

07. Corporate IT Standards **142**

CHAPTER OBJECTIVES 142

AUTHOR'S INTRODUCTION 143

SDLC STANDARDS 143

IT GOVERNANCE BOARDS 144

IT GOVERNANCE FRAMEWORKS 144

Sarbanes-Oxley (SOX) Act of 2002 145

Healthcare Related IT Legislation 145

Regulatory Impacts on IT 146

IT STANDARDS 146

BEST PRACTICES AND MATURITY PROGRAMS 147

Consistency of Outcomes 147

SDLC DOCUMENTATION REQUIREMENTS 148

CHAPTER SUMMARY 148

08. Expanding Our View of Systems **149**

CHAPTER OBJECTIVES 149

AUTHOR'S INTRODUCTION 150

SDLC TERMS AND DEFINITIONS 150

SDLC FRAMEWORKS ORGANIZE WORK 151

PHASES OR STAGES OF WORK 152

Agile SDLC Processes within an SDLC Framework 153

Multiple or Hybrid Life Cycle Paths 153

Structured SDLC Frameworks 154

Sequencing SDLC Processes 154

PROJECT PLANNING AND SCHEDULING OF WORK 155

REUSE OF METHODS AND TOOLS 156

STANDARDIZED SDLC PHASES, FORMS, AND TEMPLATES 156

BUSINESS-ORIENTED DEVELOPMENT LIFE CYCLES 157

CHAPTER SUMMARY 158

09. SDLCs Are Not Linear **159**

CHAPTER OBJECTIVES 159

AUTHOR'S INTRODUCTION 160

THE WRONG DEVELOPMENT MODEL 161

Some Early Developers Knew Better 161

DIVERGING DEVELOPMENT STRATEGIES 162

Military, ISO, and IEEE Standards 162

EVOLVING METHODS AND TOOLS 164

No Two SDLC Models or Instances are Alike 164

Start Project Planning Early 164

NON-LINEAR SDLC MODEL 165

Project Management across the SDLC 166

Concept Phase SDLC Processes 166

Parallel SDLC Processes 167

CHAPTER SUMMARY 167

10. Enterprise Change Model **168**

CHAPTER OBJECTIVES 168

AUTHOR'S INTRODUCTION 169

BUSINESS AS A SYSTEM 169

Enablers of the Business 169

A Simplified Business Transformation Model 173

Business Transformation Impact on Business Systems 174

Strategic Change Life Cycles 174

A LIFE CYCLE VIEW OF MANAGING ENTERPRISE CHANGE 175

Business Enablers 175

Enterprise Change Model (ECM) 177

ECM and SDLC Similarities 177

A MATRIX VIEW OF ENTERPRISE LIFE CYCLE PROCESSES 178

Key Concepts 178

Life Cycle Alignment 180

Dependencies Drive the Schedule of Activities 180

Project Planning and Collaboration 181

Enterprise Change Maturity 181

Defining Activities at the Intersections 182

Change Disrupts the Life Cycles of the Business Enablers 182

Business Processes Interconnect Business Enablers 183

CHAPTER SUMMARY 183

11. Basic SDLC Phase Information and Artifacts 185

CHAPTER OBJECTIVES 185

AUTHOR'S INTRODUCTION 186

BASIC DEFINITIONS FOR DESCRIBING WORK AND DELIVERABLES 186

IT ORGANIZATIONAL PROCESSES 189

Information Classifications 189

Contents

Taxonomies 189

AN SDLC TAXONOMY 191

COMMONLY USED SDLC PHASES 191

SDLC PROCESSES 192

Activity Worksheets 192

Tasks versus Work Packages 192

Project Reporting 193

Flowcharts 194

Descriptions 194

Goals and Objectives 194

Responsibility Assignment Matrix (RAM) 195

ACTIVITY (PROCESS, TASK) WORKSHEETS 195

Level of Work 196

Standards 196

Methods and Tools 197

Roles and Responsibilities 197

Deliverables 197

Work Descriptions 198

Skills 198

Resources 198

Time Durations 198

Setup 198

Quality 199

Dependencies 199

SDLC PHASE OR MILESTONE REVIEWS 200

Entry Criteria 201

Exit Criteria 201

Phase Review Checklists 201

CHAPTER SUMMARY 201

12. Integrating Disparate Scopes of Work 203

CHAPTER OBJECTIVES 203

AUTHOR'S INTRODUCTION 204

STANDARD PROJECT MANAGEMENT PRACTICES 204

Multiple Project Life Cycles 204

Project Management in Support of Agile-based Development 206

INCORPORATING AGILE CONCEPTS WITHIN A STANDARD SDLC MODEL 207

BENEFITS OF CONDUCTING AN INITIAL CONCEPT PHASE 208

PROJECT-ORIENTED WORK 208

WBS OVERVIEW 209

WBS Naming Conventions 209

Defining a Deliverables-Based WBS 209

How to Build a Proper WBS 211

How Not to Build a WBS 212

INTEGRATING PROJECT MANAGEMENT AND SDLCS 212

PROJECT MANAGEMENT PROCESSES 212

Project Management Life Cycle (PMLC) 213

SDLC versus PMLC 213

Key Personnel Roles 214

Project Management Artifacts 214

INTEGRATING PROJECT AND PRODUCT LIFE CYCLES 215

Product-Oriented Breakdown Structures 217

Software or Systems Product Breakdowns 217

Agile-based WBS 217

Contents

PMLC WBSs	219
Integrating Project Work	219
SAMPLE PROJECT MANAGEMENT WBS	**221**
PMLC WBS Template	221
PMLC WBS Template Benefits	221
SYSTEMS DEVELOPMENT WBS	**222**
Integrated PMLC - SDLC WBS Template	222
WBS Templates Must Be Tailored	222
Final comments on scheduling	223
CHAPTER SUMMARY	**223**
CONCLUSION	**225**
APPENDIX A. SAMPLE PROJECT MANAGEMENT LIFE CYCLE WBS TEMPLATE	**226**
APPENDIX B. SAMPLE PROJECT MANAGEMENT LIFE CYCLE WBS TEMPLATE	**228**
LIST OF ACRONYMS	**234**
INDEX	**238**
SUGGESTED READING	**242**
BIBLIOGRAPHY	**244**
ACKNOWLEDGMENTS	**247**
ABOUT THE AUTHOR	**248**

Figures

Figure 1.	Hierarchical Decomposition of Processes, Activities, and Tasks	34
Figure 2.	SDLC Framework as a Taxonomy to Classify and Organize Work	35
Figure 3.	Phases of the traditional 'Waterfall" SDLC Model	57
Figure 4.	Prototyping Process	64
Figure 5.	Unified Process Phases	68
Figure 6.	Big Bang model	77
Figure 7.	Modern Agile Principles	81
Figure 8.	Conceptual model of a mature DevOps environment	88
Figure 9.	Version control as part of CI	90
Figure 10.	DevOps Model with SDLC Automation	91
Figure 11.	Application of DMADV and DMAIC Improvement Techniques across a Product Life cycle	97
Figure 12.	The Cynefin framework - A sense making device	101
Figure 13.	The evolutionary development model (a.k.a. Spiral Model)	112
Figure 14.	V-Shaped Model (a.k.a. Verification & Validation Model)	113
Figure 15.	Business Strategy drives Strategic Planning, Product Conceptualization, and Program Management	134
Figure 16.	SDLC Life Cycle Phases, A Conceptual Model	152
Figure 17.	Hierarchical Structure of the Traditional Phase-based SDLC Model	153
Figure 18.	Iterative and Incremental Development Model	154
Figure 19.	Coding, Unit Testing and Integration Process Flow	155
Figure 20.	Business-oriented Life Cycle Development Model	157

Figures

Figure 21. SDLC Phase Overlaps in Context with Project Management Process Groups 165

Figure 22. Categories of Business Enablers and Underlying Corporate Assets 175

Figure 23. Enterprise Change Model (ECM) 177

Figure 24. Dependencies across the enablers. 181

Figure 25. ECM Activity Decompositions 182

Figure 26. Taxonomic View of an SDLC Phase 190

Figure 27. Finish-to-Start (FS) dependency relationship 199

Figure 28. Start-to-Finish (SF) dependency relationship 200

Figure 29. Start-to-Start (SS) dependency relationship 200

Figure 30. Finish-to-Finish (FF) dependency relationship 200

Figure 31. Splitting a Systems Development Life Cycle Across Multiple Projects 205

Figure 32 SDLC Phases Spanning Development Iterations and Releases 207

Figure 33. IT System WBS Example 211

Figure 34. Aircraft Parts Example 216

Figure 35. Passenger Jet Aircraft Exploded Parts List. 216

Figure 36. Example Software Application or System Product Breakdown Structure 217

Figure 37. Decomposition of work under Agile practices 218

Figure 38. Sample Scrum Board 219

Tables

Table 1.	Evolving thinking on traditional development concepts	56
Table 2.	Top-down, Waterfall Model—Pros and Cons	61
Table 3.	Historical references to Prototyping	62
Table 4.	Pros and cons of prototyping	65
Table 5.	Historical development of the PDSA cycle	69
Table 6.	Pros and Cons of iterative and incremental development approaches	71
Table 7.	Rad Model Tools	74
Table 8.	Pros and Cons of the RAD Model	75
Table 9.	Pros and Cons of the Big Bang Model	78
Table 10.	List of the Agile Manifesto Authors	79
Table 11.	Pros and cons of Agile development practices	80
Table 12.	Six Sigma Techniques	97
Table 13.	Data Item Descriptions (DIDs)	163
Table 14.	SDLC Phase Artifacts	191
Table 15.	Activity Worksheet Information Components	196
Table 16.	Roles of Product Managers versus Project Managers	214
Table 17.	List of Project Management Artifacts by PM process area	215

Building Our Digital World
Series Introduction

Welcome! This book is the first in a series on software and systems development practices, frameworks, methods, and tools that help us build our digital economy.

THE SERIES INCLUDES:

SDLC Foundations
An introductory and historical guide to software and systems development

Frameworks for Agility
A practical guide to achieving IT process maturity and agility in a constantly evolving industry

Tools and Templates
Improving quality and consistency of outcomes in software development and agile processes

DevOps
Communicating, collaborating, leaning, integrating and automating across development and operations processes

Welcome! This book is the first book in a series of four books on software and systems development practices, methods, tools and templates.

Over the course of this series, you will build the foundational knowledge you need to be an active and knowledgeable participant within a software or systems development program. With this foundational instruction, you will develop an essential understanding of traditional SDLC processes, business analysis and engineering practices, and modern agile concepts and methodologies. You will also come to appreciate how the values and principles of agile help overcome the problems of the traditional SDLC model, which had too often led to IT project overruns and failures.

While systems development is essentially an IT function, SDLC processes involve many people who are not in IT. IT projects are often influenced by non-IT stakeholders who may not understand the complexities and constraints faced by the IT Department. This series of books takes these concerns into account by presenting technical concepts in layman's terms to improve communications between technical and non-technical project team members and other stakeholders.

The IT industry is known for its alphabet soup of terms and acronyms. Moreover, since today's technologies quickly become the foundations for tomorrow's legacy systems, the terms and acronyms tend to collect and hang around like dust on a bookshelf. Do not be afraid! This series of books takes the mystery out of IT, and you will soon be able to communicate with the natives and grand wizards of IT.

You may be wondering why you might want to read an entire book on software and systems development practices, let alone a series of books, when there is so much information freely available on the internet. Personally, I would be the first to admit that I find the resources on the internet highly valuable. However, searches on the internet tend to be highly specific and related to some question at hand. That kind of learning, while very useful, is limited to answering questions you already know to ask.

Sure, embedded links within the returned content can lead you to a wealth of information. But it can also send you, the learner, down a rabbit's warren of contextually related information, without a guide to put the returned content into proper historical, or product lifecycle context.

This series provides a comprehensive overview of the entire subject of software and systems development, including many variant approaches, both historical and modern, in a way that is logical and easy to digest. Also, and perhaps most importantly, the Building our Digital World Series can guide organizations that need help developing or improving their corporate SDLC frameworks and process maturity.

Specifically, the purpose of this Building Our Digital World - SDLC Foundations book, and the others that follow in the series, is to help guide you through the complexities of defining, designing, developing, delivering, supporting, and maintaining software and IT systems. While this series is comprehensive in scope, individual subjects are presented at an introductory level. Some of the subjects covered within a single chapter, or even a single section, may have entire books dedicated to them. Therefore, my readers are encouraged to research and dive into more detail on the systems development processes they support. In the meantime, this series provides a comprehensive foundation to jumpstart your educational journey.

The Building our Digital World Series was originally conceived as an online training course. As a result, the information is "chunked" into small, digestible sections and subsections that deal with a specific subject within the larger chapter. The goal is to make the topics easier to discover, understand, and retain.

Throughout this series of SDLC books, at the beginning of each book and with each chapter, there is an *Author's Introduction section* where I share my experience related to the topics discussed. Through all the content I offer, my goal is to provide useful information that is factually accurate, well researched, informed by experience, and free from opinion. (Well, I'll try to limit my opinions.)

> **Life and professional experiences inform our knowledge and our opinions.**

I suspect there are IT specialists and other folks who may disagree with my opinions or have a different view on the topics presented in this series of books. Life and professional experiences inform our knowledge and our opinions, which lead to a lot of diversity in knowledge. I want to continue to learn from others as much as I hope others learn from me.

Though I have been doing this type of work for several decades, I haven't lived every experience possible in planning and managing IT projects. We learn together best through open communication and by sharing experiences and ideas. So, bring your thoughts, opinions, and recommendations to my website at www.commandresults.com. My goal is to continually increase the information on the site and to expand future editions of this series of books with relevant and up-to-date content from industry sources and input from my readers.

SCOPE OF THE SDLC SERIES

There is a wealth of information available about software and systems development. The subject of SDLC is quite large in scope, spans a host of processes and technologies, and encompasses work involving many types of skills, roles, and responsibilities. My goal is to bring a lot of seemingly disparate information together in this series, placed in proper context across software or IT based systems product life cycles, but also having each book focused on SDLC themes within a specific area of professional interest.

For example, this **Building Our Digital World - SDLC Foundations** book provides a comprehensive introduction to the evolution of SDLC concepts and approaches. As the title implies, the foundations book is a great starting place for those readers who may not have an extensive background in software or systems engineering, or for those whose understanding of systems development is limited to a narrow slice of the overall development life cycle.

The **Frameworks for Agility** book introduces a hybrid SDLC framework to manage the organization's business analysis, engineering and agile practices for software and systems development. The **Frameworks for Agility** book also includes descriptions of traditional software engineering and agile based SDLC processes and provides information on hundreds of methods, in context with the type of work performed. The process descriptions include sample information on activities, tasks, roles and responsibilities, and deliverables. Moreover, the hybrid SDLC framework presented allows an organization to treat engineering, business and agile process information as knowledge assets, and provides a structure to organize, maintain, deploy, and improve these processes over time.

The **Tools and Templates** book describes more than 100 tools and templates spanning both traditional and agile SDLC practices, plus IT project management practices. The template descriptions include usage, data and format examples making it possible to create your own templates. The Tools and *Templates* book is written in collaboration with SDLC Forms (www.sdlcforms.com), where the templates are available for purchase.

Finally, the **DevOps** book provides information on lean development strategies, plus the integration and automation of SDLC processes. Collectively, these practices streamline development activities, reduce development times, enable more frequent releases, minimize bugs and defects, and ensure development activities stay in alignment with customer and end-user needs. Specifically, the *DevOps* book includes discussions on lean development, continuous integration (CI), continuous testing (CT), and continuous delivery and deployment (CD/D) concepts and strategies. Beyond simple conceptual introductions, this book includes practical implementation strategies with the tools and technologies available today.

TARGETED READERSHIP

The primary audience for this series of books includes senior information technology (IT) specialists who may have used many SDLC models and processes over the years, as they bounced from one client engagement to another, but may not have had the opportunity to put everything together to form a complete product life cycle view to development.

Also, the careers of IT specialists are sometimes constrained by their niche areas of expertise, and those who find themselves in such situations may find this series useful to broaden their professional knowledge, and their future opportunities. Moreover, constrained by their employer's IT governance policies and standards, there are IT practitioners who may not have had much opportunity to explore and evaluate other approaches to software and systems development. This series of books will help expand their professional knowledge.

Another important audience includes IT department managers who need common-sense strategies to implement effective programs for software and systems development. AMoreover, learning how to become more agile is an important theme throughout this series, which is critical to the overall success of the IT organization to consistently build and deploy quality software applications and systems.

The books will help junior software engineers and other IT project stakeholders develop a foundational understanding of software and systems development processes, terms, tools, and best practices. This foundational knowledge provides an important baseline to understand the value of current trends to implement lean-agile and SDLC automation practices.

This series of books can help project and program managers, scrum managers and product owners, who may not have a formal education in software engineering and computer science, fill in their knowledge gaps on SDLC methodologies and processes.

The books will help executive sponsors, and other customers who pay for Information technology systems, make informed decisions on their product development and information technology investments.

Ultimately, this series of books provides practical information to anyone who has a vested interest in the successful outcome of a project or program supporting an IT product's life cycle, including:

- IT specialists
- Business managers
- Corporate and IT governance board members
- Customers
- Developers/software engineers
- End users
- Executive sponsors
- Project team members
- Process owners
- Product owners
- Scrum masters
- SQA specialists
- Testers

MAKING THE CASE FOR SDLC FRAMEWORKS AND PROCESSES

When I started my professional career, I was fortunate to work for three very large Fortune 500 companies that invested heavily in building and maintaining processes and engineering practices for developing software and IT-based systems. These SDLC process improvement programs were not one-time process documentation efforts left to stagnate over time. Improving quality and organizational efficiencies were the key objectives of their ongoing IT process maturity and improvement programs.

Each firm actively maintained and advanced their IT processes, best practices, and artifacts, in step with evolving methods, tools, and technologies. Under corporate mandates, each of the firms rigorously built and maintained SDLC processes, governance policies, and corporate standards across their information technology and IT-based product development departments. Yes, by today's standards, they followed mostly traditional development practices. But they were not afraid to embrace change as the advantages became apparent.

One of my early leadership roles involved managing a corporate program to develop and distribute reusable professional services, and related methods and tools for deployment to 4,000 business and 16,000 IT consultants who were then operating in nearly 100 countries. This time frame in my career marked a period of great change in the software industry, as the IT industry moved away from traditional SDLC practices (aka waterfall methodologies) to include less structured and lightweight approaches to software development. There was also an explosion in the technologies available to support software development. Many of these approaches are now collectively included as part of Agile-based practices.

Interestingly, I've lived through corporate transitions from traditional to agile development practices not once, but several times across my career. Even today, many organizations have not fully implemented agile practices across their systems development organizations and programs.[1]

I also supported a very large Health IT program which mandated the contractors' development teams could not use an Agile approach to development, but instead forced the teams to use the client's traditional software development practices. In other words, we had to use the client's published SDLC processes, which they believed to be more mature and practical for the size and scope of the program.[1]

My early experiences with corporate-supported IT maturity programs proved foundational across my career, as I built a deep appreciation for developing, maintaining, and evolving SDLC processes around sound agile and engineering practices. In contrast, later in my career, I worked for several organizations that did not implement rigorous SDLC processes, and it was never a fun experience. On several occasions, I was brought into situations to fix a failing IT project or program. The root cause of those failing projects almost always mapped directly back to the organization's or project team's failure to establish or follow fundamental SDLC, agile, and project management processes. Sadly, most of the problems I helped my customers resolve shouldn't have happened in the first place.

A TIME OF FLUX IN OUR INDUSTRY

Maybe the timing was coincidental, but with the advent of the dot-com era (roughly 1995 to 2001), I noticed that many of the organizations I was involved with during that time did not put a lot of rigor into developing their IT processes and best practices. Certainly, many of the startup companies spawned during the dot-com era did not have the history, culture, time, or resources to develop comprehensive IT policies, processes, standards, and procedures. These things usually come with time and through implementation of process maturity programs.

However, I also observed that many larger firms that were undergoing mergers and acquisitions began to lose step in these areas, too. In those cases, I believe the biggest driver was the inevitable loss of existing cultures, and the difficulty in rolling out IT policies and procedures, methods, and tools to the disparate companies the parent companies acquired.

There were a lot of very bright people trying to do the right things. However, change is difficult to implement and manage within an enterprise, especially when that change involves integrating people coming in from a newly acquired company. Within the IT community, we also must deal with difficulties in encouraging people to change their approach to development when they have very strong opinions on the methods and tools they have mastered and prefer.

Companies have had to deal with a great amount of change within the IT industry, including the constant introduction of new life cycle development philosophies, methods, tools, and technologies. Such change represents costs to the organization that must be justified, and sponsoring executives must balance their process improvement activities and potential benefits obtained against the costs of devoting resources to the activity. Moreover, the executives, customers (both internal and external), project team members, and other IT project stakeholders may not understand the needs or benefits of implementing new SDLC approaches and technologies. They may not be willing to take the time and effort to understand the costs and benefits of change. As occurs with most transformation initiatives within an enterprise, folks are often resistant to change even if they stand to benefit.

But I've also seen things go the other way too, where executives and IT staff are quick to jump on the latest bandwagon to implement the newest IT methodologies and technologies, or new enterprise applications, without fully appreciating the difficulties and pushback associated with implementing the proposed changes. There are almost always legacy integration or business and system transformation issues to deal with. Business processes are impacted. Stakeholders may fear negative impacts on their jobs or responsibilities. And long-term sustainability and maintainability issues must be thought through, cost justified, and addressed.

This series of books will walk you through these issues and provide the knowledge and tools you need to work effectively with your IT development teams. This series of books also provide a practical approach to build and maintain an effective SDLC framework that supports agility within your organization.

IT DIDN'T HAVE TO END THAT WAY!

I want to leave you with one final thought before I end this introduction to the *Building our Digital World* series.IT projects and programs almost never bomb explicitly due to failures by the IT department to deliver working products, or from cost or schedule overruns. Those are simple statements of the final facts, but they let the decision makers off the hook too easily for not resolving the underlying problems.

The underlying problems almost always map back to process failures, non-value-added work activities, and failures to stay in alignment with customer and stakeholder needs. This series of books provides the tools you need to diagnose the real reasons for IT project and program related failures and offers practical solutions to help turn things around. Better yet, hopefully, with this knowledge you will avoid projects failures in the first place.

01.
Key SDLC Concepts

This chapter describes key concepts for systems development that are a precursor to understanding the differences between traditional and agile SDLC processes and project management practices, plus the value of process integration across all three. In addition, this chapter provides introductory information on Hybrid Frameworks that support agility, the impact of process change, the value of using tools and templates, and right sizing your project.

CHAPTER OBJECTIVES

After reading this chapter, you should be able to discuss the following:

- Describe what the acronym SDLC stands for
- Explain the value of SDLC processes
- Explain the primary purpose of an SDLC Framework
- Explain the primary differences between traditional and agile SDLC practices
- Explain the role of project management in SDLC, and the differences when managing traditional versus agile based projects
- Explain how life cycle processes help the organization manage change
- Explain how SDLC Frameworks support IT quality and delivery objectives
- Explain how SDLC process enhances the organization's competitiveness
- Explain the importance of right-sizing software and systems development projects

AUTHOR'S INTRODUCTION

You may be wondering why I wrote this foundational book on systems development life cycle processes. The short answer is to encourage IT specialists and managers to take a systemic and life cycle view to software and systems development, from the initial product concept through its end-of- life. Moreover, this book establishes a foundational level of knowledge and semantics to enable effective communications between IT program and project team members.

> **Businesses that don't evolve, die. Evolution means change. Managing and supporting change within the enterprise is a critical element in software and systems development.**

This book introduces terms and concepts in plain English, to help guide non-technical IT project stakeholders and junior software engineers in their journey to understanding the complexities of software and systems development. Right out of the gate, this chapter helps novices get started on understanding the most basic concepts to developing and supporting software and systems.

Beyond the basics covered in this chapter, there is so much more to learn. It would be easy to pick up a book on systems development and think, "I've got it." But, not so fast. I initially conceived of this effort as a single book of about 200 pages. No matter how much I cut out, the book kept growing and growing. Ultimately, I concluded that a single book would not suffice to cover all the topics I felt were necessary to provide a sound understanding of SDLC processes. This book provides foundational knowledge to get you started before tackling the advanced subjects covered in the other three books: Frameworks for Agility, Tools and Templates, and DevOps.

As you read through this book, I want you to come away with the knowledge that this is a constantly evolving industry and that organizations must evolve with the industry to stay relevant. Why? Because IT is at the heart of almost everything that we do or use in our modern lives. With rare exception, businesses that don't evolve, die. Evolution means change. And, as you'll find in this book, managing and supporting change within the enterprise is a critical element in software and systems development.

Another important concept covered in this book is the importance of building and maturing systems development processes. I'm admittedly biased on this subject. As mentioned in the series introduction, I was fortunate to start the first decade of my career in IT, working at three different Fortune 500 type companies, all within highly technical industries. These companies made significant investments in their ongoing IT Process improvement programs, which had a big impact on my views regarding the need to build and maintain quality IT processes. Plus, I specialized as a consultant in the application of methods and tools supporting business process re-engineering and business process improvement, further driving my belief in the need to build and sustain business process maturity across the organization.

Coincidence or not, with the advent of the dot-com era, I noticed many of the companies I was involved with, either because they were new or newly formed through mergers and acquisitions, were not as disciplined. In most cases, they lacked the resources, or culture, or executive support to invest much time and effort in the development of their SDLC processes. Moreover, the enterprise-scale process and quality improvement programs I was involved in, such as Malcolm Baldrige, ISO 9000:2001, and CMMI, were such time and resource-intensive endeavors that they proved unsustainable for many organizations.

As my career progressed, I noticed that many of my peers lacked foundational knowledge on what an SDLC is, or how to read an SDLC process description or flow chart. When an organization doesn't make process documentation and process improvement a priority, the skills across the organization degrade.

I have found that many executives and project stakeholders lack the technical and historical knowledge necessary to evaluate potential development strategies and that they are quickly lost in the alphabet soup of acronyms and arcane terminology this industry constantly spews out. I also have noted a tendency to latch onto the latest acronym as a cure-all for all that ails IT. This is a dangerous practice when the decision-makers don't know what organizational changes and other investments must be made to take advantage of implementing a new development strategy, or when they do not understand the impact of their decisions on the sustainability of the organization's legacy systems.

I have written this book with two goals in mind. First, provide a comprehensive set of guides that helps a novice build the knowledge and vocabulary necessary to communicate with the IT domain experts effectively. Second, along with the rest of the books in this series, provide an SDLC Framework as a structure that helps facilitate ongoing SDLC process maturation activities without all the incumbrances of the large-scale quality and process improvement programs.

Some writers in the software industry view an SDLC Framework as a means to define and organize software development activities as a series of tasks across the life cycle of a product. Such definitions are a holdover from the heavy-weight, plan driven, and linear-sequential systems development concepts of the traditional SDLC model, otherwise known as "Waterfall." That's unfortunate, as those types of definitions miss the real point of having a framework–which is to provide structure to classify and organize process-oriented knowledge. In turn, the processes contain descriptions of best practices to help ensure consistency of work and quality across a product's life cycle.

In a modern context, an SDLC framework is a hybrid structure that incorporates full life cycle processes spanning business analysis, software and systems engineering, and agile practices. Moreover, an SDLC Framework provides the structure and infrastructure to organize, manage, discover, and retrieve information on SDLC processes.

That's my definition. You will find others. But they miss the point if they propose the primary purpose of an SDLC framework is to infer a specific sequence of systems development activities. Yes, such information is available within each SDLC process, but that is not the primary purpose of the SDLC framework. In fact, as noted in the paragraph above, that's not the purpose of an SDLC framework at all.

A hybrid SDLC Framework provides maximum flexibility, allowing development teams to choose the practices that best support the needs of their project. For example, users of the hybrid SDLC framework can choose which processes make sense for their projects, whether they follow a traditional plan-driven and linear-sequential development path, or agile. In most cases, any given IT project does not span an entire systems development life cycle, and a hybrid SDLC framework simplifies the task of selecting the processes most appropriate for the work that is funded.

Why would a modern SDLC Framework, hybrid or not, include traditional SDLC practices, you might ask? Because, like it or not, a total of 52% of the respondents in the most recent CollabNet/ VersionOne (13th) Annual State of Agile™ poll, noted that up to one-half of their development teams use Agile practices–which implies the other half, or more, are still using traditional SDLC practices. [2]

Also, agile practices traditionally do not speak much about engineering practices. Agile is about implementing efficient, iterative, rapid, collaborative, and customer-centric development processes. Agile processes live at a level of abstraction above the grunt work knowledge of how to build software products and more complex systems. Most agile processes are technology agnostic, and there is an assumption by agile methodologists that the implementing organization applies equal rigor to building their business analysis and software engineering practices. That's a bad assumption.

Now, don't get me wrong, I'm a huge fan of agile practices, However, what I think too often happens is that executive management may assume Agile is a cure-all for all that ails IT, sending the organization down yet another rabbit hole. Agile practices in and of themselves won't fix bad investment decisions, uninformed management decisions, poor planning, or bad engineering practices.

The same concern is valid for the implementation of DevOps, CI, CT, CD/D, Lean, and other modern agile-based practices. Those modern practices build upon the foundational SDLC processes. Foundational SDLC processes, particularly at the engineering level, don't go away. It's the tools and technologies that are different. Agile, software/systems engineering, and business analysis processes are all part of a three-legged stool upon which the IT department sits. And, we all know what happens when one leg is cut off of a three-legged stool.

WHAT IS AN SDLC?

SDLC is an acronym for *Software* or Systems Development Life Cycles. An SDLC defines the approach to developing software or IT systems from the initial concept through release and deployment. Fundamentally, software and IT systems development practices are highly technical, encompassing numerous engineering disciplines, skills and resources. The work is often complex and unique, and therefore largely managed under discrete projects with specific funding, goals and deliverables outlined and approved before starting work.

In a so-called traditional SDLC model, the work flows through a roughly linear-sequential series of phases, with each phase of work leading to the eventual release and delivery of a final product. The SDLC defines the processes and best practices to accomplish the work in a structured and logical manner. Traditional SDLC practices also emphasize detailed planning and documentation. The primary objectives of the traditional SDLC model are to minimize risks through rigorous planning and analysis, and to develop customer desired products within the customer approved constraints of budgets, resources, schedules and quality objectives.

Over time, a group of software engineers came to realize the application of traditional linear sequential development approaches had severe limitations that worked against the timely production of quality software. The traditional model also lacked real-time mechanisms to keep development in alignment with client needs.

Moreover, the developers were frustrated by the constraints and outdated processes imposed on them by management. These constraints introduced more risks, slowed development, made debugging a more difficult and time-consuming task, and virtually ensured the final product would come up short of meeting customer and end user needs. Worse, the developers were taking the lion's share of the blame for the failures.

The group of developers came together to define a set of values and principles, referred to as *Agile*, that help overcome the limitations of the traditional systems development model. [3] The values and principles of agile were then released in an essay titled the "Manifesto for Agile Software Development." [4]

Agile doesn't replace basic software engineering practices but is an augmentation that streamlines the execution of those processes in alignment with customer needs. Therefore, a modern SLDC preserves the organization's software engineering practices while also incorporating the processes and methods that support the values and principles of agile development.

Moreover, though agile methodologies often enforce a specific discipline for requirements analysis, it is never enough. Business analysts and product owners do not simply document business needs. They need to understand how a proposed software system must support the mission, strategy and goals of the business. In that context, they need to understand what capabilities must be delivered. And they need to understand how much the organization can afford to invest before the effort is counterproductive to the objectives of the business. Business analysis is part of the product conceptualizing process, which drives the rest of the systems development life cycle. Product conceptualization processes are part of the overall product development life cycle, and the activities cannot be ignored or thrown over the wall to IT to figure out.

Likewise, on the other end of the systems development life cycle, the product support, maintenance, and sustainability processes protect the organization's investments in products that rapidly become legacy systems. Product sustainability costs skyrocket when efforts are not taken to protect those investments through standards-based development practices and protection of assets and information through configuration management practices. Moreover, the solutions must evolve to support changing needs within the organization. In other words, the product development life cycle doesn't truly end with the release and deployment of a product.

All of these topics are covered in this foundational book on software and systems development, and then expanded upon within each book in the series.

AGILE PROCESSES AUGMENT SDLC PROCESSES

Agile practices are widely used today, but they are not in and of themselves a magic bullet. The IT specialists who developed and promoted the values and principles of agile were technically competent. They did not see technology as a constraining issue, and therefore did not spend a lot of time addressing technology beyond acknowledging the need for "technical excellence." [5]

Nevertheless, an IT department must develop and maintain sound engineering practices around the technologies and IT standards approved for use within the organization. Additionally, the agile practices must be integrated with the organization's engineering practices for either to be effective. In other words, a modern SDLC encompasses both software and systems engineering and agile practices.

The IT industry is known for its alphabet soup of terms and acronyms. And since today's technologies quickly become the foundations for tomorrow's legacy systems, the terms and acronyms tend to collect and hang around like dust on a bookshelf.

Do not be afraid! This series of books takes the mystery out of IT, and you will soon be able to communicate with the natives and grand wizards of IT.

If there is one key understanding I hope my readers will get from this foundational book, and from the other books in the Building our Digital World Series that follow, technical excellence and agility are both critical elements to successful software and systems development. One doesn't supersede the other. IT organizations make continued investments in technology, skills, resources, and processes to remain competitive or relevant. SDLC processes and engineering practices must likewise evolve in concert to support the IT department's investments in information technology and people. As they evolve, the information on processes and best practices must be captured and made available in a contextually useful manner. Continuous improvements combined with well described processes and best practices are the hallmark of a mature organization.

This series of books provides useful information and lays out a framework to achieve SDLC maturity. But none of this works if the resources and time required to develop SDLC processes are prohibitive. Therefore, an important goal of the Building our Digital World Series is to limit the negative impacts on budgets, schedules and resources to achieve and maintain SDLC maturity.

WHAT IS AN SDLC PROCESS

In a general context, the term "process" implies a series of actions taken in some coordinated fashion to achieve a desired outcome or result. In the context of an SDLC, the term *SDLC process* indicates a series of work-related activities within a systems development life cycle that are performed in a controlled manner to achieve a specified deliverable component or result or set of deliverables.

SDLC processes span all types of work in an SDLC, and all SDLC process descriptions have common elements. It's important that I describe these elements here in the beginning of this book, to avoid having to repeat all of the elements of a process over and over again. So, going forward, when you see the terms '*process*'' or '*SDLC process*' note that the following elements are implied:

SDLC Process Elements

- **Activities** (a.k.a. Summary Level Tasks)—for reporting purposes, a summary level rollup of resources, costs, and durations of underlying tasks within a process, and usually associated with a higher-level deliverable item
- **Tasks**—the lowest level of work described within a process that has a deliverable component that is both useful (e.g., adds value) and verifiable
- **Durations**—the estimated or actual time required to complete processes, activities or tasks

- **Dependencies**—establish precedent relationships to control workflow between processes, activities and tasks
- **Consumable material resources**—used and consumed within the work tasks
- **Non-consumable material resources**—necessary system components or other materials that are reusable
- **Human Resources**—labor with defined skills and experience carrying out the work tasks
- **Roles and Responsibilities**—describes the work and authorities of labor identified in the processes, activities and tasks
- **Standards**—the industry and technology standards and corporate policies governing the processes, activities and tasks
- **Methods**—a specified approach or best practice for performing work within a task
- **Tools**—equipment, software, hardware, etc., used to aid in the performance of work within a task
- **Deliverables**—the desired outcomes, tangible or intangible, of the processes, activities and tasks

In systems development, there is a hierarchy of work that flows from processes, to activities, to tasks. (See figure 1) In other words, processes are decomposed into however many activities are required to report on the work; and, activities are decomposed into lower level tasks necessary to collectively perform all the work associated with the activity.

SDLC Process elements are initially specified at the task level, and then rolled up into the activities, and ultimately to the process level. In other words, the process-level elements are built as a summary rollup of the underlying elements specified at the activity and task levels.

Conceptually, the description of all work within a process occurs at the task level; while the summary level rollups make it easier to track work efforts allocated to major deliverables of the development project.

It's important to remember that the process to task decompositions do not indicate workflow; tasks dependencies associated with the processes, activities and tasks establish workflow. Task dependencies can take the form of *inputs* or *outputs*. An input means a process, activity or task requires some other work result to complete before it can be started. An output of a process (activity or task) serves as an input to some other work-related effort.

In a project schedule, the inputs and outputs are identified as references directly back to other processes, activities or tasks. However, in actual fact, it's the deliverables that serve as the inputs and outputs between process, activities and tasks. The deliverables can take the form of tangible or intangible items, information, or results produced by the work.

Basic SDLC Concepts versus Commercialized Approaches

For the most part, this book purposely does not delve deeply into proprietary or commercial software or SDLC methods or tools. Instead, the intent is to ensure readers have a working knowledge of mainstream SDLC concepts, approaches, and strategies, and not promote or advocate the use of commercialized SDLCs, methods and tools.

There are two exceptions to this limitation on discussing highly commercialized SDLCs. First, I use Scrum as the reference model for Agile-based development practices. Scrum is an industry leading Agile methodology and is used in this book to demonstrate how Agile practices fit within an SDLC Framework. If you are not familiar with what I mean by Agile practices, hang on; the term is defined later in this chapter.

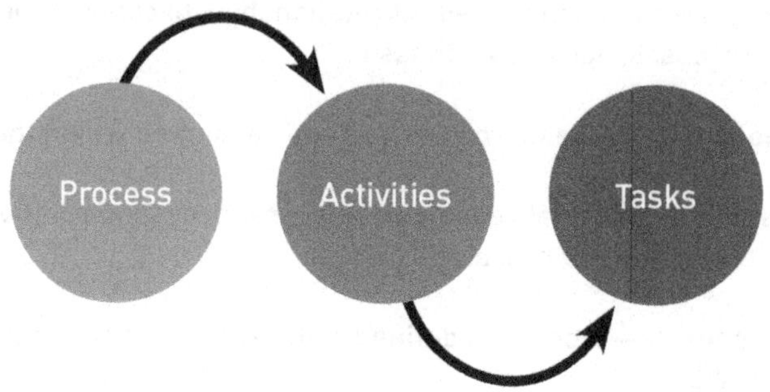

Figure 1. Hierarchical Decomposition of Processes, Activities, and Tasks

SDLC Preferences

From one organization to another, or even from project to project, different client or organizational needs can drive decisions on the implementation and use of specific SDLC processes, methods, and tools. Also, project sponsors, the development team, or IT managers may have strong preferences and promote one development approach over another.

Discussions on preferred methods and tools used by developers can be an emotional subject, given the experience and skills IT specialists have obtained over time. There is always pressure on the IT organization to adopt the preferences of the IT specialists responsible for product development, delivery, and support activities.

However, that does not necessarily have to be a problem, as there is room for multiple development strategies supported within an SDLC framework through hybrid models. The primary issue is the organization's ability to cost justify continued support of the requested methods and tools. Another issue is whether a requested method or tool adds real value over other methods and tools already approved for use by the organization's IT governance board.

IT Governance

IT governance implements a framework to ensure organizational IT investments support the strategies and objectives of the business. Having said that, corporate and IT governance programs largely evolved in response to class-action lawsuits by corporate shareholders, and to comply with resulting federal legislation, after several prominent legal cases were successfully litigated for corporate fraud and deceptive practices.

In particular, the Sarbanes-Oxley Act of 2002 was enacted in response to financial scandals involving Enron Corporation, Tyco International plc, and WorldCom. The act mandates strict rules and recordkeeping requirements on accountants, auditors, and corporate officers within publicly held companies. The act also imposes criminal penalties on those who violate securities laws. So, while the central objective of governance is to ensure corporate investments support the needs of the business, equally important is the goal to prevent corporate fraud that would open the organization and its officers to criminal liability and expensive litigation.

In many organizations, an IT governance board guides organization-wide implementation of policies, processes, technology, methods and tools. IT governance policies and standards help improve product quality, reduce the number of technologies and skills required to support the organization's IT systems and products, and ultimately help reduce product life cycle development, support, and maintenance costs.

SDLC as a Framework

Under traditional waterfall based SDLCs, there is a tendency to think of the phases or stages listed within an SDLC as representing a sequential set of processes. That was largely true at one time, but that is not the case today. We address that misconception throughout this book. For example, in chapter 2, readers learn how activities contained within the design, testing, and deployment processes must start in parallel with work initiated in the development phase.

The phases depicted within the traditional SDLC model were not wrong. The only thing wrong with the traditional model was the waterfall notion that each process flows out from the preceding process as a cascading sequence of events. However, the type of work described in the traditional waterfall model still goes on in modern Agile-based approaches. A modern approach to managing SDLC process information is to preserve the phases within the traditional SDLC model as knowledge categories encapsulated within an SDLC framework.

The concept of phases is useful as the means to organize related types of work. And, there is nothing wrong with implementing a phased based SDLC, so long as no attempts are made to imply the visual structure of the framework suggests a procedural flow of work. Instead, the phases define a taxonomy for categorizing work. A taxonomy is a system of categorization that often employs a hierarchical structure, such as shown in figure 2. The hierarchical approach implies a "*part of*" work classification relationship, as opposed to procedural workflow relationship.

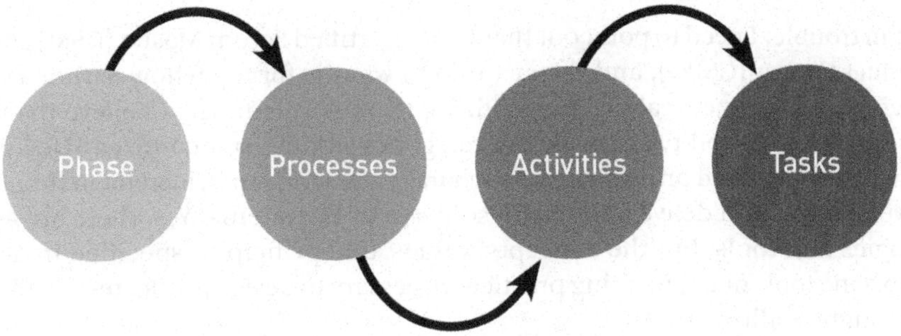

Figure 2. SDLC Framework as a Taxonomy to Classify and Organize Work

SDLCS AS KNOWLEDGE ASSETS AND INTELLECTUAL PROPERTY

The term SDLC suggests a soup to nuts approach to building, delivering and sustaining software and other IT systems. As mentioned in the previous section, it's not unusual to see an SDLC framework described as a collection of phases with related processes, activities and tasks. Each phase or stage represents a classification of similar types of work.

For example, processes, activities, methods, and tools for defining a system architecture or a product design logically fall in a design phase or stage of work. This type of hierarchical structure supported the traditional waterfall-based approach to systems development, but also created problems by implying a specific flow of work. However, it's more useful and correct to think of the collection of process related information within a phase of work as being a category of systems development knowledge. Moreover, the collection of information within a phase represents the organization's IT-based knowledge assets of best practices.

The information acquired within each SDLC phase is expensive, and time and resource intensive to collect, and should be protected as with any intellectual property. Investments in IT systems, processes, methods, and tools are justified because they enable improved business and operational efficiencies, and thereby provide a competitive advantage to the business. Having an SDLC framework makes it easier to manage, protect, and secure access to the organization's IT knowledge assets.

TRADITIONAL SDLC PHASES

The following phases (a.k.a. stages) are common in a traditional SDLC framework:

- **Concept—What do we need to build, why, and does it have a viable cost benefit?**
- **Requirements—Detailed business and user requirements analysis and documentation**

- **Design**—Development of the systems architecture and system design requirements
- **Development**—Development of the software code and system components
- **Testing**—Testing the software code and components to ensure conformance with the requirements
- **Production Turnover/ Deployment**—Documentation and transition processes to ensure that the product is ready for release to customers and end users
- **Post Deployment**—Processes to transition the solution to the IT operations teams that support users and maintain and enhance the products over their lifetime

Before I get in trouble, I need to point out that I am a Certified Scrum Master (CSM) and Certified Scrum Product Owner (CSPO), and the last thing I want is for my fellow Scrum associates to start throwing darts at me, real or imagined. So, let me explain why I believe the traditional SDLC phases still have viability as knowledge categories within a standardized SDLC framework.

Agile is a set of values and principles that encourage efficient, rapid, incremental, and customer centric development and delivery of useful software or IT systems. Yes, there are Agile-based methodologies and tools, but the concepts, values, and principles specified in Agile do not prescribe specific tools or engineering practices necessary to design, build, test, deploy, support or maintain high-quality software.

Nor does Agile prescribe specific organizational processes to conceptualize, justify, prioritize, approve, and manage IT investments. Agile is more generalized than that and is focused specifically on processes that enable efficient and customer-centric development and delivery of software and IT systems.

Each organization must build their business processes and engineering practices around the needs of the business and the technologies authorized for use within the enterprise. This is where an SDLC framework comes in handy as the means to build, organize, provide access, secure, and maintain the organization's formalized engineering practices. The SDLC framework also helps integrate IT life cycle processes with the organization's business processes.

THE AGILE SDLC MODEL

Agile-based software development practices are rapidly supplanting traditional waterfall-based development processes. At a conceptual level, Agile is not an SDLC. Rather, it is a set of values and principles that were outlined in the Agile Manifesto in 2001 by a group of leading software developers who felt strongly that traditional development life cycle approaches were ineffectual to the software development process.[6] Agile-based processes are discussed in detail in later chapters and the other SDLC books.

Across the IT industry, there are many versions and graphic depictions of both traditional and Agile-based SDLC models. Do not get hung up on the nomenclature or graphical elements of each model, as they often have more in common than not. Rather, stay focused on understanding the basic concepts, which are more universal and more important.

For now, think of Agile as a customer-centric set of processes that better support evolving customer needs, or an evolving understanding of customer needs, through collaboration, rapid prototyping, and iterative software releases. Compared to traditional SDLC models, products are released quite frequently, on the order of 1- to 4-week cycles, or even shorter durations with the implementation of advanced DevOps, CI, CT, and CD/D capabilities. Most importantly, under Agile, each new development iteration of a product adds incremental value to its customers and end users.

An Agile-based SDLC enables cross-functional teams to work in parallel on a set of iterative development processes that look like the traditional SDLC processes noted above, to include:

- **Planning**—Determining development priorities and focus for the upcoming release

- **Requirements Analysis—A deeper dive into understanding user requirements**
- **Design—Expanding the software or system design to accommodate the new functionality**
- **Coding—Developing the software code that implements the new set of requirements**
- **Unit & Integration Testing–Ensuring that the new software code works as intended and that the new code does not break other software codes or components under development**
- **Client Reviews—Ensuring that the new iteration of software meets customer expectations for the features and functionality implemented, and obtaining feedback and customer insights for future improvements**
- **Release—Deploying a working product that is potentially releasable to the customer and end users**

Agile as a concept does not describe specific processes used to complete work within each development life cycle iteration. Likewise, Agile at a conceptual level does not recommend specific methods, tools, and technologies to perform software development work. The values and principles of Agile are agnostic to methods, tools, and technologies.

Yes, there are commercialized Agile-based methods and tools, and those offerings often prescribe specific methods and tools. However, Agile as a set of values and principles is agnostic.

Moreover, descriptions of a typical and generalized Agile-based development iteration cycle do not get into the details of development, testing, and delivery related work. For example, regression testing, performance testing, provisioning and deployment practices are not listed as part of the traditional Agile model, although they can be added as work items within an iteration of work (sometimes referred to as a "Sprint") or included as part of the processes associated with an Agile-based product release.

By the way, this is also true of traditional waterfall based SDLC models. Both traditional and Agile models are conceptualized and presented at a high level. It's up to organizations and their project teams to specify the details of how these activities are performed, and what tools are authorized.

TRADITIONAL SDLC MODELS VERSUS AGILE PRACTICES

No book on SDLCs can be complete without discussing traditional waterfall-based development life cycles and contrasting the traditional SDLC model with modern Agile development practices, such as the Scrum-based Agile framework. In this book, I discuss the differences and pros and cons of each approach to software development.

Most importantly, I also introduce SDLC framework concepts, such as the hybrid SDLC Model, which allows an organization to implement the best of both traditional and Agile SDLC practices. Hybrid SDLC approaches allow project teams to tailor their life cycle development activities to meet the business, user, support, and delivery preferences of their executive sponsors and customers.

AGILE AS AN SDLC IS NOT ALL-INCLUSIVE

Organizations contemplating a change to Agile-based practices should understand that Agile processes have a focus on implementing methods for improving work team interactions, developing software that meets customer needs through frequent customer collaboration, and improved responsiveness to changing requirements. Agile, as noted, does not address software engineering technologies necessary to build, test, deploy and sustain IT products and services across their life cycles.

Agile-based practices make up only a small part of the IT Organization's SDLC. The IT Organizations must continue to develop engineering processes and approved practices in support of the authorized technologies approved by the IT Governance board. An SDLC Framework, as proposed in this book, provides the means to build and maintain a library of the IT Organization's complete life cycle development processes, including both agile and engineering processes.

What's in an SDLC Process?

By definition, a process is an assembly or sequence of activities that collectively achieve a specific result. For our purposes, we'll define the terms activity and task to have the same meaning as work. E.g., mental and physical actions that result in creating a deliverable result of some type. Moreover, a process is a set of activities and tasks that collectively achieve a higher-level deliverable result.

We can't simply list the activities and tasks associated with a process and be done with it. We need to know the sequence of the activities and tasks. SDLC processes are prescriptive, defining sequences of work within their descriptions via their process, activity and task dependencies.

We also need to know something about the resources, skills, equipment, and tools required to complete each activity. Assuming a schedule of activities is also required, we'll need to know the estimated duration time for each activity and the precedence relationships between the activities.

Those who are new to performing the work might like to see a description of the work and expected deliverables (a.k.a. work products) associated with each task. Moreover, they may require detailed instructions on how to perform the work.

In this series of SDLC books, I introduce an *Activity Worksheet* that makes it relatively easy to document work requirements. Subject matter experts complete the activity worksheets during an SDLC maturity documentation effort or as part of an initial project planning exercise. Activity worksheets document work requirements and can be used at a process, activity, or task level. Ideally, an IT organization maintains their activity worksheets within an electronic repository and organized within the constructs of the classification structures defined within their SDLC Framework.

Protecting Organizational IT Assets

Failure to maintain system components and related information artifacts can cause additional work, project delays, and cost overruns when a new project team has to take over and fix any problems that crop up over time. I cannot tell you how often I have taken on a software enhancement project only to discover that the organization has not preserved the original code, system components, test plans, test and production environment configurations, test scripts, or product support information. When that happens, the costs and time required to enhance or modify the software increase dramatically. In some cases, it may be impossible to completely replicate the original artifacts, test environments, and product functionality.

The application of configuration management and version control processes to manage configuration items are critical to the long-term management of organizational IT assets.

> **Configuration management (CM)-**is both a governance and a systems engineering process that supports the management of components, documentation, information, and other artifacts related to an IT system. CM processes help the IT organization sustain their IT systems over time by maintaining an accurate record and store of all components and artifacts associated with each release of a system or systems.
>
> **Version control-**is a component of CM that assigns a unique revision identification number to track changes to documents, computer code and programs, website content, and other IT product-related artifacts. Documentation includes requirements analysis, architecture, design guides and systems administration guides, training aids, program and project related

communications, and deployment and transition plans. Note that the version control identifier is unique to the artifact it is attached to and says nothing about how it supports the system as a whole.

Configuration item (CI)-refers to any item related to a software product or IT system that might change over time. CIs include requirements, design and architecture specifications, software code, information and data models, workflow descriptions, work products, other related artifacts, and support and user documentation. CIs also include network devices, licensing, key documents, hardware, IoT type devices (or anything that may be internet aware); literally, anything that's included in the makeup of the system, or related documentation and training aids.

CM processes and tools enable IT organizations to identify and track individual CIs and document their purpose or functional capabilities and dependencies with other CIs. The information maintained on CIs, usually within a CM repository, documents which versions of IT components and other related information artifacts, go together with each release of a software product or system.

Moreover, data associated with each CI lets the IT organization know if and how a change in one system CI impacts other CIs associated with that system. Collectively, the implementation of version control, CIs, and CM processes and tools enable restoration, maintenance, enhancements, and debugging of software and other IT components over the life cycle of a product or system.

PROJECT PLANNING AND MANAGEMENT

Since projects, by definition, are complex and unique work endeavors, they are subject to some degree of uncertainty. It should be obvious that project planning must start before a project is kicked off. However, given the uncertainty and improved understanding of project requirements over time, project planning is also an iterative task throughout the project.

This statement is true for both traditional waterfall and Agile-based projects. The only real difference is that an Agile-based development team and their assigned product owner can easily change product delivery priorities on an ongoing basis. For example, under Agile, product owners and the development team make product features and function development decisions through a series of ongoing sprint reviews and backlog refinement meetings throughout the project's duration. In the traditional model, any change within the project must be approved by the customer or executive sponsor via a formal change management process.

Still, as with traditional waterfall approaches, the iterative work planned by the Agile-based product development teams should be documented and managed within the project manager's updated project plans, change logs, and schedules. Project plans, change logs, and schedules provide the link back to the customer's charter that specified and approved the deliverables and constraints of the project.

Project Integration and Tailoring

Since the scope of work varies across projects, each involving a different set of processes and activities, project integration and tailoring are two of the most important roles a project manager has. Project integration is an activity whereby the project managers combine processes and activities from different disciplines into a project schedule. Project tailoring involves modifying processes and activities to fit project needs.

As an example of project tailoring, project managers must collaborate with their project teams to plan and execute the right set of work within the specified constraints of the project (e.g., scope, budget, resources, schedule, quality). A given project seldom encompasses the entire set of processes defined within a corporate SDLC. The project manager and team members must determine which set of SDLC processes fall within the authorized scope of work.

As an example of project process integration, project managers have a set of processes and best practices to manage project-oriented work that are integrated with the business analysis, product conceptualization and engineering-oriented work. All processes that are relevant to the needs of the project must be noted in the project plan and integrated into the project schedules.

Another example of project process integration involves combining processes and activities to support changes in business processes that will be affected by new software or system changes. In those cases, the project schedule might include processes for training and rolling out the IT solution and related business process changes to end users. In addition, the project plan and schedules must include the processes that define the organization's approved deployment of the product to customers or end-users, and the support and maintenance organizations.

Project Management in an Agile World

At the beginning of this section, I said that the work planned by Agile-based product development teams is documented and managed within a project manager's updated project plans and schedules. Some Agile practitioners may disagree with that statement. In some cases, they may believe that the discipline of project management is unnecessary to the successful delivery of Agile-based projects. However, bear with me for a minute while I explain why the discipline of project management is still necessary under Agile-based development practices.

Agile practices evolved to address many of the issues that the sequential and lengthy waterfall development processes introduced. Agile was not introduced as a reaction against project management, per se, but rather an indictment of the traditional SDLC practices and corporate dictates that are unrealistic. Software developers knew there were better and more efficient ways to develop software.

To help prove my point, note that it was a group of leading Agile software developers who introduced the Agile Manifesto.[7] Within the Values and Principles of the Agile Manifesto, the fourth stated value reads:

> **"Responding to change over following a plan."**

This might sound like an indictment against project plans. However, note the follow-on statement:

> **"That is, while there is value in the items on the right, we value the items on the left more."**

The writers and proponents of the Values and Principles outlined in the Agile Manifesto rebel against enforced corporate processes that are unnecessary, don't add value, are inefficient, and don't help the customer. They also rebel against dictates from management and customers that are unrealistic given the scope of work that must be accomplished with the time, budgets, and resources made available to complete the work.

In contrast, the theories behind program and project management practices evolved from a management perspective, well before information technology was a thing. Most simply put, the primary goal of project management is to deliver a unique product, service, or result within the client's or executive's approved constraints of scope, time, budget and quality.

Project management practices stand apart from an organization's systems development processes and their corporate/business processes. Project management practices exist to help a project team plan the work required, and then initiate, monitor, and report on the progress of their work in terms of achieving the goals and objectives established and approved within the project's Charter. Project management reporting provides visibility and directs management attention to address any problems when a project is in trouble.

While developers may be quite motivated to build high-quality and customer-centric products efficiently, paying customers still want someone making sure the project team stays focused on delivering the features and functions that support the organization's strategic goals and objectives, and completion, within the constraints that justified the investments. Product owners and scrum managers are focused on development priorities and delivery issues. In contrast, the project manager's job is to ensure the project stays in alignment with the charter and constraints that justified the investments.

Bottom line, Agile projects are just as susceptible to being shut down if the paying client does not believe the development team can deliver a useful and useable product within the project's approved constraints. Moreover, paying clients want an independent project manager paying attention and reporting on the team's progress against the project's scope, time, budget and quality metrics.

As a final note on this subject: Project managers under the traditional waterfall based SDLC model can exert influence and control that is antithetical to developing high-quality and useful products promptly. Loosening up that control is ultimately a major part of Agile. Project managers must adjust their thinking to see that their role is to plan, monitor, and report progress against the approved metrics and the planned release deliverables, and not dictate day-to-day development activities and priorities.

LIFE CYCLE METHODS ARE FOR MANAGING CHANGE

Projects are unique and temporary endeavors that have uncertainty regarding scope, the best development approach, risks, and expected outcomes. In contrast, operations-oriented development work is continuous, repetitive, seeks to promote consistency in both processes and deliverables, and minimizes the opportunity for change—especially unexpected change, or risk.

Therefore, it is not a stretch to state that an underlying theme in project management is managing the process of change; or, more precisely, ensuring that the organization has available and is using proper project management processes, methods, and tools to manage change. SDLC processes in a mature DevOps environment with continuous integration and delivery capabilities, can look operational. However, even then, the products developed are still unique and novel. In either case, the discipline of project management has a role to play in managing change within the organization, which is a topic I discuss in more detail in chapter 3.

For now, bear in mind that project management life cycle (PMLC) processes include best practices for planning and delivering work and project deliverables in the face of uncertainty. In contrast, mature SDLC processes help reduce the risk of uncertainty in project-oriented development work by operationalizing the product development tasks.

An SDLC documents the organization's best engineering practices, methods, and tools for creating the intended IT product, service, or result of the project. An SDLC documents all the information related to product development and categorizes it into manageable pieces to ensure successful outcomes. In a sense, the SDLC decomposes product development work into manageable processes, activities, and tasks - spanning requirements gathering and analysis, design, development, testing, and deployment and into the operational support and sustainment phases of a product's life cycle.

Enterprise Change Model (ECM)

IT is heavily intertwined in virtually everything a business does today. Moreover, cross-functional and cross-departmental processes are facilitated through IT systems and applications. A change in one business process very likely impacts other business processes. The IT department must be able to look at the organization holistically to build and sustain relevant and useful business systems.

Commercial businesses, government entities, and nonprofit organizations all must evolve to stay relevant, and evolution requires change. With this understanding, it should be apparent that developing and sustaining best practices for life cycle product development management are critical success factors for an entity's long-term survival. This book introduces a high-level enterprise change model (ECM) that depicts relationships between the internal and external factors that drive the business, the resources that enable the business, and life cycle processes that facilitate change across the enablers of the business.

Examples of business drivers may be new competitive threats, or new customer and product opportunities, or changes required by government policies and directives. Within this book, we broadly classify enterprise resources as customers, products, services, customer-oriented solutions, human resources, equipment, and delivery and supply-chain partners. The life cycle processes contained within the ECM are the SDLC and PMLC processes the organization implements.

Corporate Strategy Drives Change, Realized via Programs and Projects

In the previous subsections, I have discussed the need for organizations to develop life cycle processes for both project management and systems development to support the organization's change initiatives. However, what drives the need for change? The short answer is that change is driven by internal and external factors that present opportunities, or impediments to the organizations' ability to realize its mission.

For an organization to succeed, executives must ensure that their businesses stay aligned with the needs of customers and prospective customers. In this customer-oriented context, the development of a business strategy is the means by which the executives determine what must change within the organization to ensure that the business is properly aligned to achieve the mission, goals, and objectives of the business.

Portfolio investment decisions are defined and prioritized based on corporate strategies, goals, and objectives. Organizations implement portfolio investment decisions via programs and projects. While strategic change may require the development of new products and services, it's equally possible that these changes may impact business processes, systems, people, and other organizational assets, all of which make up the "business system."

Specifically, this book presents some of the business drivers that drive organizations to change, and why it is of critical importance that they approach change from a systemic view. In other words, the organization must look across all the enablers of a business to determine the scope of change required to implement a new corporate strategy and have the resources in place to coordinate and manage change across the organization.

Thus, the allocation of funds and human resources to develop and enforce standard approaches to life cycle systems development is critical to achieving the strategic objectives of the enterprise. The same is true for developing project management life cycle (PMLC) processes, which are discussed in this and other books within the *Building Our Digital World* series.

An Enterprise View to Change

From the previous section, it should be clear that strategic change encompasses so much more than developing new software, or even—in a broader sense–developing new IT-based solutions. Rather, strategic change initiatives may impact all the *enablers of the business*.

At a conceptual level, the enablers of business include its people, processes, information, technology, customers and the organization's products and services. At a more granular level, people as a business enabler include employees, contractors, and other human resources and partners that the organization can rely on to support its mission, such as supply chain and delivery partners, resellers, and channel partners.

Process business enablers encompass the business processes that help the organization efficiently support its business operations. For example, the organization may develop business processes to support distribution, finance, manufacturing, marketing, procurement, sales, services, supply chain, and transportation functions. Process enablers also include the project management and SDLC processes mentioned in this book.

Information business enablers include the organization's intellectual property; but also include documentation, reports, information, and raw data. IT is one of the critical enablers of the business, which includes computer hardware, software, storage, security, and networking components. However, IT business enablers also include existing and emergent technologies that enable manufacturing, transportation, distribution, warehousing, and other critical business functions.

Perhaps the most important business enablers are the organization's customers and prospective customers, and the products and services they require. If the organization does not have a clear idea of who their target market customers are and can't produce the goods and services their customers want, they are on their way out of business.

EXPANDED VIEW OF A SYSTEM

The need to support and manage enterprise strategic change initiatives across the enablers of the business creates an expanded view of what a system is. Business processes and their related information systems facilitate business integration, across the enablers of the business, to support the mission of the enterprise. With this complexity in mind, change cannot occur within one business enabler without potentially impacting other parts of the business.

In other words, a business enterprise is a complex system of interconnected resources that require life cycle management. In that context, the IT products developed by the organization are just one of the organizational resources that require life cycle management. Both business - and product-oriented systems - are related and collectively support the strategic goals and objectives of the organization.

As an example, let's use an SDLC for a new automobile. New car models come out all the time due to changes in consumer desires, technologies, and competitive forces. Imagine what it would be like trying to design, build, and deliver a modern car without knowing anything about the target market customers and their requirements, what they can afford to pay, and the overall design objectives of the car. Those market and financial considerations lead to the design and development of the body, chassis, engine, drivetrain, brakes, steering, cooling, heating and air conditioning, entertainment, navigation, and supporting computing and electrical systems. Moreover, there are dependencies between the components that impact design, development, and test decisions.

In our automobile life cycle example, various organizations, partners, and suppliers manage the design, build, test, and delivery considerations for which they are responsible. They may use any number of integrated business processes to coordinate their activities. So, in effect, the tasks involved operate as a complex and integrated system.

An Enterprise Life cycle Change Management Problem

In the previous automobile analogy, note that we were not speaking of the operational aspects of building individual cars for the customers. Rather, our focus was on the life cycle change aspects associated with bringing a new car to market. The primary point to be made, in this rather simple automotive systems example, is that strategic change initiatives impact systems beyond the products and services provided by the organization. In this context, implementing strategic change can be viewed as an enterprise-class life cycle management problem. More importantly, an organization needs standard methods and tools for managing the development of complex systems of all types.

MULTIPLE SDLC PATHS

Over time, different ideas have emerged on best practices to develop customer-focused software, systems, and solutions. All have their relative strengths and weaknesses, and different SDLC approaches may be appropriate from one project to another. I mentioned in previous sections how a hybrid approach to developing an SDLC framework could accommodate more than one SDLC approach.

SDLC decision criteria are driven by how much change is expected or allowed in the requirements over time. The amount of control management and customers desire to exercise over the project affects the execution of processes within an SDLC. Moreover, the amount of time and effort customers and users are willing and able to devote during the development life cycle impacts SDLC process decisions.

It is important for corporate IT organizations to develop, publish and train their staff on a set of best SDLC practices to help minimize project risks and provide better consistency in providing project deliverables on time and budget. The corporate SDLC may include multiple or hybrid approaches to address different project needs. Moreover, IT organizations must maintain SDLC maturity programs to improve development and operational efficiencies.

Integrating PMLC and SDLC Processes

When the project managers come into a program or project, they bring with them skills for managing the delivery of work that is non-operational. They are experts in managing work that involves an element of risk, such as:

- Nonroutine work
- Work where the plans and scope of work are subject to change
- Work where there is some degree of uncertainty regarding product requirements

- **Unique work that has a non-tested development approach**
- **Work that has undetermined risks**

At the start of the project, if not before, the project manager develops a project plan, working with the project team and various *Subject Matter Experts* (SMEs). That plan includes a detailed assessment and scheduling of the work that will be required to create the deliverables. While the project manager brings skills for managing the project, the technology, and subject/domain experts determine the approach needed to create the deliverables of the project. The processes for both project management and systems development must be integrated, and that activity starts with the development of the project's work breakdown structure (WBS).

SDLCS FOR COMPETITIVE ADVANTAGE

It is unrealistic to think that different organizations must have the same SDLC processes for managing their development projects. For example, an organization can implement agile practices and technology for greater efficiencies and responsiveness, and thereby obtain competitive advantage.

Using a corporate SDLC for competitive advantage does not mean it cannot employ industry standard processes. Often, simple tweaking of existing SDLC processes, implementation of reusable tools and templates, and documentation of best practices can have a big impact on performance. Also, there are any number of commercial and open SDLC methodologies that an organization can build on.

SDLC processes require maturation, as with any critical business process within the enterprise. Process maturity supports consistent use of best practices to help ensure high quality and efficiencies across IT programs and projects. Likewise, reusable tools, templates and forms are useful to capture and deploy the organization's best practices, the development of which should be part of every SDLC maturity program. The topic of SDLC maturity programs is introduced in the next section.

SDLC PROCESS IMPROVEMENTS AND MATURITY

Well managed organizations have developed process improvement programs to document and then incrementally improve their critical business processes. The goal is to improve operational efficiencies through continuous improvement of their business processes. Improved operational efficiencies are realized when the organization decreases development and deployment times, increases the volume of product outflows (i.e., throughput), and reduces costs.

Organizations can choose from alternative process maturity models, such as continuous process improvement, capability maturity, or quality assurance models. Continuous process improvement programs seek incremental change and associated benefits by minimizing project risks. Process improvement programs also minimize time and costs on IT projects.

Capability maturity reflects the degree to which an organization maintains a formalized approach to documenting and improving their business processes. Within the IT industry, and especially by federal contractors working on federal IT projects or programs, the capability maturity model integration (CMMI) is an example of a widely used CMM.

QA is an approach taken by an organization to define, measure, and improve quality. Quality is an important consideration when assessing both processes and products. For example, how do we know what "good" looks like if we haven't first established acceptance criteria that are clearly articulated with measurable performance standards? Within the IT community, it would be remiss for a client's proposals and contracts to not include *acceptance criteria* as quality objectives that state what is acceptable for deliverable items.

There is a downside to implementing QA, process improvement, and capability maturity programs. First and foremost, these activities represent a cost to the organization, and most organizations are not in business to develop and sustain these programs.

In my experience, it's not unusual for such programs to fail from the start, or lag over time. It takes time, effort, and resource commitments to initiate and sustain these programs. Management needs to communicate the benefits and encourage participation. However, it's equally important to take steps to make these programs as unobtrusive as possible. Executive management and process owners must audit and encourage participation on an ongoing basis to achieve and maintain the benefits.

While implementation of QA, process improvement, and capability maturity programs can be resource- and time-intensive, this book provides a relatively unobtrusive approach, using activity worksheets to document and then improve SDLC processes over time. Activity worksheets define work at the process, activity, or task level. Since the type of information collected is similar, we don't need a different type of form or template for each application—unless you want to. To minimize confusion, I have elected to simply call the template an activity worksheet for all instances of use. The type of work documented in the activity worksheet (e.g., process, activity, or task) is identified in the template.

The activity worksheets for SDLC processes and related activities and tasks are categorized in the context of an organization's standard SDLC framework, and the documents are usually managed within an electronic document management or portal software system.

REUSABLE SDLC TEMPLATES

Consistency in the application of best SDLC practices across the enterprise leads to better software, better systems, and better solutions. One of the best ways to achieve consistency is to encapsulate best practices into a set of reusable forms and templates and then ensure widespread distribution and use of these artifacts within each of the organization's systems development projects.

For this Building Our Digital World - SDLC Foundations book, Command Results has partnered with Documentation Consultants.[8] They produce professional quality SDLC templates that have been used and vetted across thousands of customers, spanning small, mid-size, large, and Fortune 500 companies, as well as government agencies, nonprofits, and consulting firms.

In Modules 2 and 3, we describe the use of more than 100 reusable templates in context with the SDLC and PMLC processes where they apply. With the information provided in those two modules, users gain the knowledge they need to appropriately select and tailor the templates that best apply to the needs of their projects.

It is important to note that most projects require only a subset of the templates described. Some templates apply more to traditional development approaches, while others support Agile and project management practices. The project manager, working with their project team, customers, and sponsors, select only those templates that are most useful, without overly burdening the project team with unneeded activities and documentation.

RIGHT-SIZE YOUR SDLC IMPLEMENTATION

As we have noted in the previous subsections, most businesses need to change from time to time, and each new change requirement necessitates executive sponsorship, financial and human resources, IT support, and life cycle management. Change is inevitable no matter the size of the organization. But on the positive side, systems development and PMLC processes can reduce risks and improve the consistency of outcomes in all organizations. Moreover, a robust SDLC approach should be flexible enough to be used across all types of business, product, systems and services development needs.

Having said this, we do not expect that a "one size fits all" approach works well when implementing an SDLC framework.

An organization's size and resource constraints dictate the scale of an SDLC process management framework they can develop and sustain. Each organization must tailor their SDLCs processes, methods, and tools to ensure an appropriate level of oversight and governance, while not shutting down the business in the process. Adopting and applying industry standards and using activity worksheets to document best practices within the organization can help streamline the overall effort.

CHAPTER SUMMARY

This introductory chapter covered basic SDLC concepts, and the reasons corporations should invest in developing mature SDLC models and processes. There are two prevalent software development models. One is the traditional plan-driven and linear sequential view to development, often referred to as the Waterfall approach. The other approach to development implements the values and principles of Agile, as outlined in The Agile Manifesto.

Modern software and systems development frameworks may implement a hybrid approach that allow aggregation of the organization's best business analysis, engineering and agile practices. Similarly, a corporate SDLC framework may facilitate deployment of multiple development approaches or hybrid development models. A hybrid SDLC strategy offers more flexibility to development teams and customers who can select the best development approach and set of practices that support their unique project needs.

Corporate strategy drives both product and organizational change. The discipline of project management helps the organization manage change by separating out the unique and usually time-critical work into dedicated programs and projects. Those programs and projects may prefer to use either traditional or agile practices, or some form of hybrid approach, to plan and complete their work.

Strategic change initiatives within an enterprise encompasses more than changing products and services. A large-scale change initiative also affects all the enablers of the business, such as people, processes, information, technology, products and the customers the organization supports. As a result, organizations need to take a systemic view to understanding both the scope of work and impact to the business and its enablers.

SDLC frameworks provide structure and organization to corporate knowledge on software and systems development, support, and sustainment processes across the lifecycle of a product. Rather than reinvent the wheel on every project, so to speak, a process maturity program helps the organization document, deploy and improve their business analysis, engineering and development processes on a continuing basis.

The organization must be taken into consideration the amount of work and time dedicated to the effort. Though useful, at some point there are diminishing returns in the effort applied to process documentation and improvement activities. This book recommends that the organization's best practices can be encapsulated and deployed via reusable processes, tools and templates.

Under the deterministic approach of the traditional software development model, project managers integrate and tailor project schedules by pulling processes and practices into a detailed project plan and schedule. Reusable forms, tools, and templates aid in the planning and scheduling of work.

Agile approaches tend to take a more empirical view to software and systems development. Instead of predefining all requirements and development tasks across the life of a project, the developers build the product through a series of short iterations with each new cycle creating a new increment of functionality. The developers evaluate work for an upcoming cycle based on current customer requirements and feedback on the functionality released in previous increments. Still, the development team can select appropriate business analysis and engineering practices from the SDLC Framework, as opposed to completely defining a new approach during every iteration.

02.
Evolving Life Cycle Development Approaches

This chapter describes foundational practices that most modern SDLC methodologies employ to some degree or another. Not all the listed approaches are practiced in every SDLC methodology, but each will include one or more. Understanding these basic development concepts helps project team members and other completely defining a new approach during every iteration.

Formalized methodologies help project stakeholders quickly come up to speed when introduced to new software development approaches. This chapter presents the six approaches in sufficient detail to help readers understand their historic role, applications, and value.

CHAPTER OBJECTIVES

After reading this chapter, you should be able to discuss the following:

- Describe the alternate descriptions and definitions of the SDLC acronym
- Describe the evolution of SDLC approaches over time
- Explain the primary differences between the following foundational systems development Life cycle concepts:
- Traditional, sequential SDLC processes (waterfall)
- Prototyping
- Iterative and Incremental Development (IID)
- RAD
- Big Bang
- Agile

AUTHOR'S INTRODUCTION

I have personally managed software development projects going back to the late 1980s. For the first 15 years or so, the project teams I managed pretty much followed traditional waterfall-based processes. Then, I began to support teams that preferred using early variants of extreme programming (XP), prototyping, and other agile-based approaches, such as Scrum. Don't worry if you don't understand what these terms mean; you will by the end of this chapter.

The Agile-based projects I supported, invariably, had a significantly faster turnaround on the release of new product enhancements that our team could deliver to our customers and end users. Our customers enjoyed viewing the prototypes we delivered for their review and gave us valuable feedback on how the product could be further improved. The shorter life cycles and faster turnaround on the release of new software products reduced the risk of producing bad code, reduced the time to debug and fix the code, and reduced the time wasted on development activities that were not useful or valued. Let me explain.

First, I'll address the issue of risk. In a traditional product development life cycle, the preponderance of the developer's time is spent on requirements gathering, analysis, and product design, testing, and debugging. The time spent on coding is relatively minimal. For example, I have seen estimates of the developer's time spent on coding that range between 10% to 50%. These estimates are consistent with my observations.

In the traditional model, most testing activities do not occur until the completion of the requirements, design, and coding activities. By that time, the team may be months into an annual development cycle. When bugs are found by testers, as they inevitably are, it becomes a complex and lengthy task to debug the code to find all the bugs, fix them, and then test the code again. This sequential development approach increases the risk of bugs, and results in schedule and cost overruns.

Another risk under traditional waterfall-based development strategies is the length of time it takes to produce a deliverable. Often, product releases were scheduled around an annual budget, to align with the corporate budgets that authorized the project. A lot can change in a year. It was not uncommon for one of my project teams to deliver a product and then hear customers and end users exclaim something to the effect of: "Cool product; can it do this or that?" Under an annual product release cycle, the only answer is: "Yes, it could do that... however, you never told me that's what you wanted. We can put that requirement into next year's product release."

Of course, that's not a very satisfactory answer. In fact, it's a terrible response, but it's the only reply possible under traditional waterfall-based development practices. Not only do requirements change over the course of a year, the end users often don't know what's possible until they can see and use the software or system. Their eyes become open to the possibilities, as they can visualize process and data relationships they may not have considered before. That's frustrating to the development team, as most developers want to deliver a product their customers value and use.

In contrast, Agile-based development practices incorporate prototyping, and IID approaches help solve these problems. However, to be successful, up-front and ongoing input from customers and end users is vital. The development team members can't make up requirements for their customers. Customers must tell the developers what functions they need the product to provide. So then, as product releases are built and delivered, the developers need client and end-user input to further inform priorities and client needs for new features, functions, and enhancements.

The biggest challenge I've experienced with the introduction of Agile-based development practices is to get customers to commit their time and that of their employees to requirements gathering sessions and iterative product reviews. Often, customers don't know how to explain what they need, or how to express their needs, and they may not believe they have time to devote to product reviews and requirements-gathering sessions. Executives, product owners, and project managers need to stress the importance of participation in these activities. Also, the development team brings methods and tools that aid in understanding, gathering, and documenting customer requirements.

In the Chapter Objectives section, I list six foundational software and systems development concepts. There are other development approaches, many of which are discussed in this book.

However, these six concepts serve as foundational strategies that form the basis of many modern software development processes and methodologies.

During my research for this book, I saw that some writers lump other development approaches into the category of SDLCs, such as Feature Driven Development (FDD), Component-Based Software Development, Object-oriented (OO) Development, Joint Application Development (JAD), and so-called lightweight methodologies (such as Crystal Methods). Most of these examples offer enhanced approaches to software development. However, they are not foundational concepts or full-life cycle development approaches, and therefore should be included as potential processes and related activities maintained within an SDLC framework.

Others who write about SDLCs methodologies and practices may include discussions on commercial offerings, such as IBM's Rational Unified Process (RUP)®. While I don't have any issues with such commercialized offerings, it is not the goal of this book to promote or otherwise be an advocate for commercial product offerings. My primary goal is to stay focused on presenting and explaining foundational concepts. However, I can't ignore them either, so brief introductions are made to some of the more significant advancements on the fundamental SDLC concepts.

The last topic covered in this chapter is an introductory discussion on process improvement, collaboration, and automation methods that can enhance the Agile model. What's fascinating to me is that I helped implement several of these productivity improvement disciplines, such as Kanban and Just in Time (JIT), when I started my career working as a project manager in a high-tech manufacturing facility. These concepts initially evolved to help improve efficiencies and lower costs, rework, and inventories in high-throughput manufacturing environments. Today, they are being used to help improve efficiencies in software and systems development, by introducing product feature enhancements as they are needed, and reducing the amount of work in process (WIP) that complicates the work, increases the opportunity for system-level bugs, and reduces developer productivity.

The stated principle within the Agile Manifesto has proved challenging to achieve: "Agile processes promote sustainable development.[9] The sponsors, developers, and users should be able to maintain a constant pace indefinitely." The emphasis on sustainable iterative development and delivery practices makes developer burnout a real and constant concern. SDLC process automation, the IT analog to business process automation, offers a real solution to achieving the sustainability principle. SDLC process automation capabilities are implemented as part of the DevOps, CI, and CD/D concepts described in this book, and then again in much more detail in my book on SDLC Modern Practices.

SDLC AS AN ACRONYM

Through the years, the term SDLC has morphed from a focus on software (e.g., software development life cycles) to an expanded view of systems development life cycles. In either case, the acronym is used across the IT industry to describe several life cycle approaches to developing software and more complex IT solutions. In some organizations, the SDLC acronym refers to solutions development life cycle, which emphasizes that the whole point of the activity is to create a product, service, or result that solves one or more business or customer problems.

In the strictest sense, an SDLC describes a set of phase-based processes required to develop software or IT-based systems. These phases traditionally include initial concept definition, requirements gathering and analysis, design, development, testing, deployment, and sustainment. We must be careful, though, in how we think of the term *phase*.

While the word "phase" can imply a distinct stage of a product development life cycle, we should not think that the ordering of phases implies a specific ordering of work within an SDLC. Rather, we need to think of phases as simply classifications of work that logically go together. In this context, the phases listed within an SDLC serve as a taxonomy of work required to develop and sustain software and IT-based products.

In this book, I define SDLCs as a framework containing the organization's processes and methodologies that help the organization develop customer-focused IT solutions efficiently, and with repeatable outcomes. In this context, the term SDLC applies not only to software application development but also to the development of complex IT-based systems of virtually any type.

THE EVOLUTION OF SDLCS

This book is not meant to provide a history of computers, programming languages, or branded SDLC methods. Rather, the focus of this book is to help my readers understand and apply essential yet common SDLC concepts, practices, processes, methods, and tools. However, some historical review is necessary to help my readers understand how and why software and systems development approaches evolved from traditional, top-down approaches—which provided structure but also came with some degree of rigidness, to modern Agile concepts and practices that offer more flexibility.

Specifically, this book provides historical context to show how selected SDLC methods evolved to address deficiencies in traditional development methods, and to take advantage of emerging methods and tools that better support evolving business needs and uncertainty in requirements. I do not advocate for one SDLC approach over another, nor will I spend time reviewing proprietary methods and tools. Those can be highly emotional subjects that are, quite frankly, better addressed by an organization though its IT governance program.

In the Resource section of this book, I provide URLs to online articles plus referenced books for those readers who want to delve more deeply into the historical development of software and systems development concepts.

Applications and Technologies Drove the Evolution

Modern electronic computing systems have a fairly recent history, with the first practical electronic computing systems introduced in the late 1930s. In the earliest days of computer programming, the only models for developing complex and expensive systems came out of the construction and manufacturing industries. At the time, it seemed logical that the plan-driven, linear-sequential approaches used in those industries would equally apply to developing computing systems.

What's important to understand is that during the first three decades, computer systems development was extremely complex and expensive, with early programs typically focused on solving complex math, science, and engineering problems. Such problems could be solved in a plan-driven, linear-sequential approach, as early programming methods and computing equipment put a premium on getting things right the first time. As an example, this author is old enough to have experienced a situation where a developer ran simulation code on a legacy mainframe computer at night, and the program went into a continuous do-loop and cost thousands of dollars in charge-back fees with no useful result.

In the traditional development model, a developer would first need to understand the customer's requirements, then architect and design a solution, then develop the product, and finally test and deploy or turn over the new product. This development process logically progressed sequentially from start to finish, with only some overlap for planning and preparation between the major phases of the development effort.

Strengths of the Traditional Development Model

The traditional plan-driven, linear-sequential development model would not be around still today if it didn't provide some benefits. I would go further and say the traditional model is still preferred when major changes to a product's features and functions are impractical to incorporate once development has begun.

For example, and by way of analogy, you are not likely to see agile methodologies implemented in a large-scale fashion by shipbuilders, or builders of skyscrapers and large-scale utilities or infrastructure projects. The cost of large-scale changes becomes increasingly expensive the longer the project goes on. The same issue applies to building large-scale enterprise software applications. Once the basic architecture and designs are locked in, it's very expensive and time-consuming to make wholesale changes to the basic behaviors of the application and its navigation and usability features.

There are other areas where the traditional model works well, such as:

- **When the customer mandates through contractual obligations the development of specific capabilities, features, and functionality: › In such**

cases, a rigorous plan must be created and followed within the approved constraints of the project's charter

- **When specific capabilities, features, and functionality are documented as part of an initial business analysis and concept phase of work: › Initial project planning is enhanced, making it easier to monitor and report progress against the original plans**
    - The architecture and designs require fewer costly and time-consuming changes during development
    - Customer and end-user participation are limited to supporting initial requirements gathering sessions, user acceptance testing, and periodic attendance in milestone or gate reviews and final product approvals.
- **When it's possible to break a product into independent components, it's possible to stagger the use of resources across the component development activities to achieve better utilization.**
    - Also facilitates concurrent development of components and construction of integration interfaces against an established architecture and design
    - Established requirements driving rigorous architecture and design assessments before development can help improve component interoperability, separate system functionality and eliminate redundancy, simplify the design, and enhance overall system performance.

Limitations with Traditional Development Methods

As noted in the previous section, plan based and linear development approaches work well when fabricating buildings, ships, railroad lines, freeways, or other large infrastructure and utility projects. In those cases, it's very difficult to make major design changes after development has started, and those systems are expected to work, without major changes, for decades.

Given the expense, complexity and scope of work involved in developing software, it seemed logical to both project managers and engineers that they too should implement a plan driven and linear-sequential life cycle approach to development. But that didn't work out so well. Instead, the software development industry suffered through a high number of project failures while attempting to follow the traditional life cycle development model.

Confounding the problem was the tremendous growth in computer power and software development tools, making software applications increasingly large and complex. Traditional SDLC approaches introduced new problems in this environment, and it became apparent within the IT industry that the linear-sequential development approaches have critical limitations. For example:

- **Customers do not always know their needs up front, or even what's possible.**
- **Product enhancements or customer reviews of the software are available only once a year or so.**
- **The traditional approach does not support changing an established plan to address a changed requirement or to support continuously evolving requirements and needs.**
- **The linear-sequential process holds off testing until the application is completely coded, making it extremely difficult, time consuming and expensive to find and fix any bugs or defects.**

- Without periodic customer reviews and opportunities for realignment with customer needs and expectations, the chances of catastrophic failures are greater.

On the positive side, from the 1970s through the 1990s, there was an explosion in software development methods and tools that made other approaches to developing software possible, and in fact, necessary. This book explores the new software development enhancements from a fundamental's capabilities perspective, beginning with the very next section.

Modern Software Development Methods and Tools

This chapter discusses six foundational development concepts that have matured over the last several decades, and in some cases from the mid-1950s. Though some of these terms may seem dated, their underlying concepts persist as the basis for many modern software and systems development methodologies.

In direct contrast to the traditional linear-sequential development practices, modern programming methods and tools make it practical to build software incrementally through any number of relatively short-term development iterations. This enables more frequent releases of useful functionality, as opposed to building a product all at once before releasing anything

The incremental development practices also make it easier for development teams to accommodate change requests and to allow the product to evolve in support of customers' changing requirements. Such accommodations are the hallmark of agile-based development practices, where customers and end-users can review the product and make new recommendations prior to each development iteration. Ongoing product demos and reviews help guide further development of the product, as each iterative update provides clients and end users with new insights on how the product can be used.

Some of the development methods and tools that support modern, Agile-based practices include:

- OO programming
- GUI development tools
- Prototyping
- Iterative and incremental development techniques
- RAD tools
- Model-driven development capabilities

Don't worry if you don't know what these things are. I'll explain them all in later sections within this book. For now, you should understand that these development methods and tools make the following benefits possible:

- Development of customer-centric software
- Incremental improvement in the product's value, in shorter duration iterative development cycles
- Improved development efficiencies from finding bugs and other problems earlier
- Frequent opportunities for customer and end-user reviews and redirection
- Ability to adapt software to address changing or previously unidentified requirements

Incremental, Iterative, and Prototyping Approaches

Before I get to the heart of this section, let's introduce a few key terms:

> **Prototype** is a limited use application, or sometimes an existing application or system that can serve as a conceptual model to both evaluate usefulness and areas for improvement.
>
> **Iteration** implies repeating short-term development cycles, which typically include activities for requirements gathering and analysis, design, coding, and testing.
>
> **Incremental** describes the evolution of the product, as each new development iteration adds value by expanding the product's capabilities, features, and functions.

IID and prototyping are the most important development concepts to emerge. Collectively, these development methods address many of the shortcomings cited for traditional plan-driven, linear-sequential development methods. The benefits of prototyping and IID include:

- Prototypes are inexpensive and relatively quick development builds or installations of commercial-off-the-shelf (COTS) software.
- Prototypes allow customers and end users to see what an end product might look like, how it could be improved, and how new features and functions support their business processes and enhance usability.
- IID cycles make it possible to consider implementing modifications to support changing requirements sooner than with traditional approaches.
- IID delivers value to the organization in shorter time scales, offering more capabilities more quickly.
- Discover inadequate architectures and poor designs before changes become expensive, if not cost and time-prohibitive to implement.

Notice that I have not mentioned agile as a practice in this list. The reason for this, as discussed previously in the Author's Introduction section, is that Agile is a collection of values and principles, and not specific methods, tools, or techniques.[10] Nevertheless, in actual practice, agility is achieved through a blend of IID and prototyping practices, along with using development tools that make requirements gathering and analysis, design, coding and testing more expedient and efficient.

Software Development Focus Shifts to Systems Development

As computer science technologies, methods, and tools improved, the size, capabilities and attendant complexity of software systems increased. Also, the increasing complexity of IT systems, including hardware, networks, storage, data integration tools, and messaging brokers, plus Internet- and Intranet-based services, makes the notion of the SDLCs acronym applying only to software as rather limiting.

Software is now an integral part of most everything, including business systems, games, automobiles, the home, appliances, robotics systems, aircraft, and Internet-based trading hubs and communities. As a result, SDLC as an acronym has more commonly become known as a systems development life cycle or even a solutions development life cycle.

Regardless of the name applied to the SDLC acronym, organizations need a common set of best practices for managing the development life cycle process. As with any business process, the IT organization should continually improve and mature their SDLC processes.

Using Frameworks to Support SDLC Maturity

As with other organizational processes, SDLC processes are of no value unless they are documented, used, and improved over time. To have any value, corporate business processes of all types must be identified, organized, documented, maintained, distributed, discoverable, and improved. These seven elements are the essence of process maturity.

- **Identified**—How do we conduct our business?
- **Organized**—How is the process information classified and structured? (e.g., a taxonomy)
- **Documented**—What is the organization's or industry's best practices?
- **Maintained**—Where does the organization maintain its process information?
- **Distributed**—How does the organization allow authorized users to access the process information?
- **Discoverable**—How do authorized individuals' access and find process information when they need it?
- **Improved**—How does the organization improve their processes over time?

Many corporations implement a standardized SDLC frameworks and set of processes as part of their IT governance activities. Those frameworks provide the necessary infrastructure to support an ongoing SDLC maturity program. If your organization hasn't established an SDLC Framework, this book, along with the *Frameworks for Agility* book, describes a standardized SDLC framework that can support an SDLC maturity program.

FOUNDATIONAL SDLC PROCESSES

This section introduces software development approaches that form the underlying foundations of most software development methodologies used today. At a conceptual level, the following six approaches hit all the basics.

- Traditional, sequential (waterfall)
- Prototyping
- IID
- RAD
- Big Bang
- Agile

This list is certainly not representative of every software development method in existence, as the focus is on foundational concepts, and not specific methodologies. This chapter describes these six software development approaches in some detail.

If you are a project manager, you may be called upon to manage projects across any number of SDLC methods during your career. With this in mind, this book intends to provide enough familiarity to understand the underlying approaches to development, some historical context, and the strengths and weaknesses of each software development approach.

Understanding these six basic concepts will give you a leg up in understanding formalized SDLC methodologies your employer may prefer. Still, when a project team member or other stakeholder becomes engaged in a software or systems development project, and they are unfamiliar with the approach proposed by the organization, I would encourage them to seek out additional information and training on the specific development methodology chosen for their project.

TRADITIONAL/ WATERFALL SDLC MODEL

The traditional software and systems development model (a.k.a. Waterfall) is a highly structured approach that has its roots in the manufacturing and construction of large, complex, and unique items. The waterfall metaphor has a conceptual basis that development life cycle work proceeds sequentially and linearly, with little or no opportunity to revisit preceding activities or variation in steps, with the output of one process flowing into the next. Planning and project management are the hallmarks of the traditional development model for large-scale, complex and unique systems.

The linear and sequential life cycle development approach has been around for thousands of years. The approach works particularly well on large, complex, and expensive construction projects. For example, when constructing a ship, large buildings, industrial sites, utility plants, or roads and freeways, the budgets, plans, requirements, and designs are completed up front, before development starts. The upfront planning and design activities are necessary because there is little opportunity to implement major design changes once development begins.

The waterfall approach is practical for many industries, particularly where major design changes are expensive if not impossible to make once a project starts. For example, imagine the impact of changing a ship's design to include 10 new decks or another 100 feet in the span after the ship is already under construction. Those things can be done, technically speaking, but in most cases it's probably not practical nor economical.

In the early days of software development, given the engineering complexities, technology limitations, and scope of work involved, it seemed sensible to follow the traditional plan-driven and sequential-linear approach to development. Likewise, given the uniqueness of the work, the discipline of project management was enforced to control costs and schedules, and to help guide the development team to achieve the scope of work that justified the time, expense, and resources allotted by management.

For the most part, this is the way software was developed from the 1960s through the early 1990s. Table 1 documents some of the people and organizations that contributed to the use of the traditional model for software and systems development.

Traditional Waterfall Model—Covering the Basics

The traditional model is easily one of the earliest and better-known software development methodologies. It is a highly structured approach for building large-scale systems where requirements are well known and not likely to change. Moreover, the waterfall approach works well where there is no practical or cost-effective way to backtrack and rebuild parts of the solution, or incrementally and iteratively design and build the solution.

By analogy, one cannot imagine building a skyscraper without developing the architectural plans for the building's entire structure before starting construction. Nor would a builder want to tear down some part or all of the skyscraper in the middle of construction, to redesign and build a new tower because the customer's requirements changed. As another example, it is hard to imagine contractors who are building a freeway from San Diego to Atlanta, suddenly rerouting the Freeway through Chicago when they are already halfway through Kansas.

Again, could these things be done? Yes, but not without much greater costs, additional resources, and significant time delays. It's not likely the benefits of such changes would outweigh the costs.

Waterfall is not Dead—Yet

Despite rumors to the contrary, the waterfall approach is not dead—for one simple reason: no other approach gives senior management confidence, even if sometimes misguided, that their investments are progressing as planned, and headed toward the desired and expected outcome.

Table 1. Evolving thinking on traditional development concepts

Who	When	Why
Herbert D. Bennington	1956, 1983	Initially termed the "Stagewise Model," Bennington described the sequential software engineering phases used in the SAGE Program; however, Bennington later republished the article with a forward stating that the phase-based processes were not sequential, and that they also extensively practiced prototyping.[10]
Winston W. Royce	1970	First to formalize the waterfall process; but he also made the following comment: "In my experience, the simpler method has never worked on large software development efforts."
Thomas E. Bell and T. A. Thayer	1976	First to coin the term waterfall.[11]
Department of Defense (DoD)	1985, 1994	Military standard DOD-STD-2167 formalized the waterfall method in 1985.[12]; later superseded by MIL-STD-498[13]
Learmonth & Burchett Management Systems (LBMS), under contract to the Central Computer and Telecommunications Agency, a UK government office that was later subsumed by the Office of Government Commerce (OGC)	1981 - 2000	The structured systems analysis and design method (SSADM) is an example of a waterfall-based approach to development that incorporated structured analysis (SA) and structured design (SD) methods

Under traditional waterfall practices, managers and sponsors can verify completion of planned deliverable artifacts during the product development life cycle, such as requirements documents, design specifications, prototype (as an option), code, test results, and deployment plans—all in line with the project's charter, budget, and schedule forecasts. The plan-driven approach to development makes it easy to see if the project is still on plan, or not. If not, actions can be taken to bring the project back in line with the plan, or, with customer approval, the plan can be changed.

Waterfall Model Process Overview

Software projects initiated under waterfall typically span one fiscal year, although there is nothing that prevents an organization from authorizing and funding longer or shorter software projects.

Under waterfall, the objective is to complete all development life cycle processes in one iteration. For that reason, waterfall typically involves a lot of planning, requirements analysis, and design assessments before starting the development and testing processes.

Traditional SDLC phases (Figure 3) include Concept, Requirements, Design, Development, Testing, Deployment and Sustainment. Each Phase includes one or more processes that depict work associated with that phase of work as a set of activities and tasks. The activities and tasks specified in the waterfall based SDLC model include information on the corporate standard engineering practices, methods, and tools.

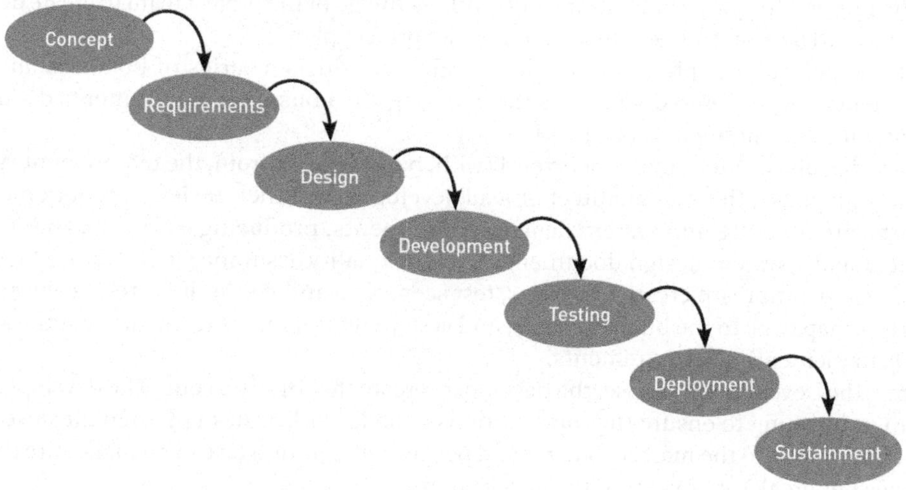

Figure 3. Phases of the traditional 'Waterfall" SDLC Model

Industry Standards governing the Traditional Model

The nomenclature and terms used within the waterfall model vary across organizational implementations. DOD-STD-2167A - Defense Systems Software Development, published February 29, 1988), is the only standard to formalize the traditional waterfall-based lifecycle approach. DOD-STD-2167A was superseded by MIL-STD-498, on December 5, 1994. MIL-STD-498 moved beyond describing processes for software development, to define life cycle processes appropriate to developing more complex systems.

The newer MIL-STD-498 standard was extremely prescriptive in terms of requiring the development of detailed *data item descriptions (DIDs)* for each phase of the development lifecycle. DIDs are documents that fully describe the data deliverables required by DoD contractors at each phase or stage of the product's lifecycle. DOD is an acronym for the Department of Defense.

MIL-STD-498 defined and formalized project management practices, supported the use of computer-aided systems engineering (CASE) tools, and allowed the execution of overlapping and iterative SDLC processes on non-safety-critical projects. In other words, MIL-STD-498 allowed flexibility to use CASE tools and iterative practices to reduce development cycles times, even as the DIDs and project planning activities created additional work across a development project.

The DoD standards are no longer supported, but the original DoD standards served as a baseline for developing the ISO/IEC/IEEE International Standard for systems and software engineering. The current international standards are the *ISO/IEC/IEEE 15288:2015 standard for Systems and software engineering -- System life cycle processes*, and the *IEEE 12207-2017 - International Standard - Systems and software engineering -- Software life cycle processes.*

Traditional Waterfall Phases

I'm going to leave the industry standards, as they have moved away from defining strict guidance on specific SDLC process methodologies and the flow of work. Instead of using the terms *phase* or *stage* to categorize types of work, the international standards use the term *technical processes*. There is nothing wrong with the technical process descriptions defined in the industry standards, but here in this section I need to show both the type and logical flow of work.

For this and all three books in the series, I will use the Phase-based terminology identified in Figure 3. Traditional SDLC waterfall phases. It's fine if you prefer to use the term technical processes. Regardless of which terms you prefer (e.g., phase, stage or technical process), the important thing to know is that in a modern context they all refer to collections of common types of SDLC related work.

In the Concept phase, business analysts and other stakeholders create business plans, perform costs analyses, conduct concept of operations (CONOPS) investigations to determine high-level business and user requirements, and finalize the program and project charters. The concept phase is the best time to initiate the project management planning process. During the project planning process, the project manager works with members of the project team to create detailed project plans. The resulting documentation is the project plan.

In the Requirements phase, a business analyst conducts a series of business and user requirements sessions and documents their findings in a business requirements document (BRD) and user requirements specifications.

The Design phase often runs concurrently with, but staggered from, the requirements phase. In the design phase, the lead architect or lead developer start their review of requirements to define the architecture and system design requirements, producing a systems architecture document and a system design document. Also, the quality assurance (QA)-testing members of the development team create test plans, test scenarios, and test scripts. Test scenarios and test scripts map back to the business, user, and system requirements to ensure that the system properly implements all requirements.

During the Development phase, the developers create the software code. The developers also perform unit testing to ensure that new sections of code work as desired, then the developers link their new code to the master code set and perform integration testing to make sure the new code does not break other parts of the application.

The Testing phase presents a bit of an anomaly in the waterfall model. As noted in the Design phase discussion, preparation for testing starts in the Design phase with the development of test plans, test scripts, and test scenarios. Also, the Development phase includes unit and integration testing. However, there is a lot more testing left to be done in a software development cycle.

In the waterfall model, most testing is completed after developing the software application code. For example, the development team may perform regression testing, systems testing (e.g., end-to-end testing), user acceptance testing, and performance testing.

Regression tests are defined and created to ensure that a new system or changes to an existing system can work on the same dataset and return the same results. Regression tests help ensure that changes to a system do not break existing functionality.

End users run through the test scenarios and test scripts during user acceptance testing (UAT). Under waterfall practices, this is the first-time end users interact with the product.

For large applications, performance tests ensure that the system meets desired performance metrics for responsiveness, scalability, reliability, and concurrent usage.

Already we can see that traditional waterfall practices depart from implementing truly sequential phases. Testing phases processes, activities, and tasks run concurrently from the Design phase through the Testing phase.

The Deployment phase (a.k.a. Production turnover phase) in the traditional waterfall model includes deployment management and transition planning processes. The deployment management process helps ensure that the receiving organizations (e.g., customer support, technical support, and IT operations) are prepared to manage the software application or system upon release.

The Sustainment phase includes operational support processes and potential product enhancements. Product support includes bug fixes, customer and end-user support services (e.g., help desk), and system support and administration duties.

As with processes within the testing phase, deployment phase processes start back in the requirements and design phase and continue throughout the product's life cycle. Deployment processes direct the development of the documents necessary to support a product after release (e.g., system admin guides, technical support guides, database administration guides, customer support guides, training aids, business process documentation updates, transition plans, and deployment plans.)

Once a product is released, the integrity of the product must be maintained. Product-sustaining activities are operational, and may include system and database administration duties, software configuration, installing software enhancements, software reinstallations, and installing security updates.

Note that the need for new product enhancements often triggers a request for funding and approval to proceed with an entire development life cycle under the traditional software development model.

Benefits of Traditional Waterfall Concepts

First and foremost, traditional waterfall approaches have a heavy emphasis on creating, maintaining, and deploying sound engineering practices. Agile as a concept does not address technology or engineering disciplines necessary to design, build, and test software products and systems. Agile does not dictate processes for maturation of corporate SDLC processes around the technologies approved or the documentation of best practices used within the organization.

Under the traditional waterfall model, although admittedly burdensome, all the planning and delivery documentation helps the IT organization sustain their corporate IT investments. Large-scale enterprise systems undergo continuous changes to stave off competitive pressure and support the implementation of evolving corporate strategies and end-user needs. The documentation produced helps the IT organization sustain these critical and expensive IT assets across their useful life cycles.

Another reason why waterfall approaches are still popular has to do with limited access to customers and end users. It is often a challenge to obtain a commitment from customers and end users to participate in an initial requirements phase. In contrast, many Agile methods require much greater time commitments from customers and end users throughout the development life cycle.

Finally, waterfall approaches have matured, and hybridized versions include prototyping and even iterative development cycles. So, there's nothing that says the waterfall approach cannot be tailored to adjust to project needs and project risks. In my book *Frameworks for Agility*, I show how IT organizations can integrate structured SDLC processes with prototyping and IID processes.

Waterfall Model—Uses

The waterfall model for software and IT systems development is most practical when the application requirements are well-defined and not subject to change. Waterfall-based projects best support ongoing legacy systems development projects, where the technology may not support iterative development practices, or the investment is so large and complex that there is no easy way for developers to backtrack on the architecture and functionality previously implemented.

The waterfall model works well when customers devote time and access to their domain experts at the beginning of the project, and when the requirements are not subject to a great deal of change once the development phase begins. Because requirements drive the entire waterfall method, all the way through deployment, it is critical that the project team gets it right the first time.

This is not to say change is not allowed. Some change is inevitable in a project-based environment where uncertainty is a key characteristic. However, change must be controlled and approved through a formal change management process. Failure to do so invites a critical review by the project's sponsors. If the sponsors believe the project is out of control or no longer worth the budget, time, or resources, or unable to deliver key features or functionality, they might cut their losses and terminate the project.

Waterfall Model—Scope Creep

When requirements are not properly defined upfront, especially in a waterfall development project, there is often tremendous pressure coming from the customers and user community, directed to the software development teams, to add in missed requirements. The same thing happens when the project manager allows the introduction of new requirements in a waterfall-based project. Invariably, that leads to busted budgets and missed delivery dates. In short, we cannot change the scope in the waterfall model without impacting the other project constraints of quality, schedule, resources, and budgets.

Here's the problem: miss a delivery date or bust a budget on a large enterprise application development project and the project team may have more attention than they want. Moreover, the project team may lose the support of paying customers or executive sponsors if they cannot get the project back under control. The IT organization must implement a rigorous change management process to ensure that the paying customer agrees to any modified or expanded work scope before starting the new work.

Waterfall Model—Defining the Requirements

It should be apparent by now that the number one challenge when using the waterfall approach to software development is requirements gathering. The development team is expected to understand all the requirements before completing the design and the start of coding. With this in mind, the waterfall approach works best when the customer and end users are willing to spend time at the beginning of the project specifying their requirements, and the requirements are not expected to change.

Usually, a business analyst is assigned to help document business and end-user requirements. The system architect and other members of the development team need to work in parallel to analyze the requirements and create an architecture and system design that can deliver the specified capabilities.

While there is no set rule, within a year-long software release cycle it's not unusual to devote 8 to 12 weeks to documenting user and system design requirements before starting development. This time frame would encompass two to three phases, including a requirements phase, a design phase, and possibly prototyping phase. In some cases, in parallel with the requirements analysis, it may be possible to start building the foundational architecture components for the solution.

In a prototyping phase, business analysts document initial requirements through a series of requirements-gathering sessions with domain SMEs. To aid in the collection, the project manager, business analyst(s) and systems architects may schedule a series of collaborative workshops in the form of JAD sessions.

The development team often creates a GUI or mockup screens to show proposed workflows, navigation, logical data manipulations, and data field concepts. The domain experts review these visual aids to look for areas for improvement and refine the prototype accordingly. Also, when necessary, test data and workflow capabilities are implemented in the prototype to demonstrate support for impacted business processes.

Waterfall Model—Solution Applications

The waterfall approach to software development is best suited for situations where the requirements and technical approach are well understood and not subject to change. The waterfall model is a highly structured approach that has well-defined processes that are easy to monitor, control, and use to gauge progress against the approved project plan. The waterfall approach is often preferred by executive sponsors who want monitoring and controls to ensure that the project team is delivering within approved scope, quality, time, and budget constraints.

Another concern that waterfall addresses is getting the design right up front. The reality of software development is that there are multiple ways to do almost anything, but not every approach is appropriate for the needs of the client. For example, the business and end user requirements impact which design will be the best approach to support performance, security, scalability, and usability objectives.

If the development team has a sound understanding of the requirements at the beginning of the project, they have a better chance of creating the right architectures and designs to support those requirements. Getting it right the first time means avoiding time-consuming and costly rework at later stages of the product's development life cycle.

Waterfall Model - Challenges

I have noted a couple of times now that one of the challenges with waterfall is that it is seldom easy to obtain or discover all the requirements up front without spending a significant amount of time on detailed requirements-gathering and analysis efforts. Even then, it may be impossible to capture all the requirements up front, or the requirements may change throughout the project.

As an example, I led a 2.5-month-long logical solutions design documentation effort to refine and understand the business, process, and technical implementation issues surrounding eight application modification requests specified by the customer's executive sponsors. The solution requirements analysis team consisted of a business analyst, both domain and business process experts, and several application development specialists. The business and end-user requirements involved implementing several out-of-the-box data fields on a COTS-based application, implementing new data relationships in several tabs within the end-user application, and integrating other external business applications to populate revised data fields.

Figuring out how to implement just eight customer requirements may not sound like much work, especially since each business requirement only took one or two paragraphs to document. However, the technical implementation issues led to a little more than 70 separate development work orders to describe the features, functions, and design changes that the development team would have to implement to support those eight requirements. The business processes also had to be modified to accommodate the system changes, some driven by the desired requirements and some driven by the necessary technology changes. Things are seldom as easy as they seem.

This requirements analysis effort occurred on a large, multi-year development project using a hybrid (waterfall-Agile) SDLC model that also employed continuous and parallel development strategies. The business requirements were documented in the initial startup phase of the project, while the development team focused on implementing foundational architectural changes to the underlying business applications.

After the initial requirements phase, the desired product changes were prioritized, scheduled, developed and released in a series of monthly iterative development cycles over the remainder of the fiscal year. Iterative customer reviews and tweaking were allowed to some extent, but major change requests were put into a queue for consideration in the next fiscal year's budget.

With this example in mind, waterfall and hybrid models only work when the end user customer is willing to invest the necessary time up front to help define all the requirements for the release and they are willing to wait, sometimes a year or more, to see the final results. Moreover, the customers and end users must be able to live without introducing major enhancement or modification requests until the current and funded development project is complete.

Table 2 summarizes the pros and cons of the top-down waterfall method.

Table 2. Top-down, Waterfall Model—Pros and Cons

Pros	Cons
Better suited for large enterprise and legacy systems applications development, where development languages or the complexity and size of applications make it difficult to rebuild major functionality	Until the customers see the software, they may not understand the totality of their needs, leading to costly redesign, development, and testing efforts
A highly structured and managed approach that emphasizes documentation to demonstrate compliance with requirements and authorized project constraints	Lacks the flexibility to adapt as requirements change or become better understood
Works best where requirements are well understood and not subject to much change	The lengthy amount of time from project initiation until delivery, and the inability to consider new changes in requirements
Customer and end-user involvement are limited to the requirements and testing phases	The requirements have to be right the first time
Documentation requirements provide detailed records on charter, requirements, architecture, design, code, testing, and deployment decisions and artifacts	The design may not easily accommodate changes once the customers and end users finally have access to the product
Continuous reporting against the project plan makes it easy to see progress against the authorized scope, budget, and schedule constraints	Customers need adequate time up front to support requirements gathering and analysis activities

Waterfall Model—Summary

The traditional waterfall-based approach to development is the earliest documented approach to software development, modeled after practices used in the construction, utilities, and manufacturing industries. It is best suited for projects where the requirements are well-known and not subject to much, if any, change. Executive managers and customers may prefer this approach, as they can monitor progress against a rigid plan and set of requirements, project budget, and time constraints. Also, the traditional waterfall approach may be more practical when the paying customer cannot commit resources to support the ongoing iterative reviews that are required under Agile-based development approaches.

On the other hand, software developers began to realize that the software development process has a different set of attributes and drivers than projects that deal with developing buildings, roads, ships, or other large and complex industrial development projects. Most critically, it is challenging to discover all customer and end-user requirements before the end users have a chance to see and work with a proposed application. Moreover, the competitive demands of business can drive rapid change, and the supporting business systems need to evolve to meet those demands quickly. Prototyping and IID better suit those situations.

In the subsections that follow, I am going to visit the prototyping model, IID model, RAD model, and Big Bang models, all of which were developed to overcome the limitations of the structured waterfall approach to software development. This chapter concludes with a discussion of Agile as being a set of values and principles that encourage project team interactivity, efficient software development practices, customer collaboration, and responsiveness to change.

PROTOTYPING MODEL

A prototype is an incomplete version of a proposed product that is used to demonstrate concepts to customers and end users before developing the final solution. Prototype software development is a practical component of both waterfall and Agile approaches.

It may seem that a prototyping cycle is very similar to the IID scenario where multiple development cycles are involved in building production software before final release. However, the key differentiator is that a prototype may be completely different than the final product and should be considered a throw-away effort.

The value of a prototype is that it "simulates" an approach to solving a problem and may validate a design approach to creating the final solution. Prototypes serve as a discussion point between customers, end users, and developers on how the system can be enhanced to support business and user needs, thus helping refine the functional requirements. As a design tool, a prototype can help validate a technical approach to providing a set of features and functions that satisfy the needs of customers and end users while also meeting performance, security, and reliability requirements.

Prototyping is not a complete SDLC, put is typically part of both traditional and agile development practices. Prototyping may span several development iterations within the hybrid waterfall models and various Agile and iterative and incremental approaches. Prototyping has been around and described in historical papers, such as those shown in Table 3.

Table 3. Historical references to Prototyping

Who	When	Why
Herbert D. Bennington	1956/ 1983	Bennington notes that the approach they used in 1956 was dependent on a prototype
Barry Boehm	1986	Prototyping and iterative development are major components of the Spiral Model, described in Boehm's paper titled: "A Spiral Model of Software Development and Enhancement."

Covering the Basics

One of the issues associated with a traditional waterfall approach is that customers and end users do not always have a complete vision of what capabilities they will need from a system. Lacking a complete vision or understanding of requirements is one of the main concerns with enforcing a waterfall-based development approach. Without a prototyping phase, it may be impossible to get a clear and complete definition of requirements from customers and end users ahead of development. Also, developers may need to test different design approaches before deciding on a final architecture that can meet all the demands of the users.

Prototyping offers a potential solution to these dilemmas. Business analysts initially gather the requirements from customers and end users, and the development team then creates a limited prototype that the customers and end users can see and use to some extent. After reviews by the customers and end users, the prototype may need to be enhanced or rebuilt to test other concepts.

In short, prototypes offer a cost-effective means to test ideas and get customer input before making final development and design commitments. The use of prototypes almost always leads to a better understanding of the customer, business, end user, architecture, and design needs.

A software prototype can have varying degrees of complexity. At its simplest, the prototype may include some mock screenshots with a description of the proposed menu options, data fields, and underlying workflow. More sophisticated development organizations, particularly those that have implemented model-driven development capabilities, may be able to go further and implement working software with functional menus, data structures, and workflow. Moreover, there's nothing that prevents a prototype from being used as the first release in an IID project.

Prototype software development is useful for risk mitigation when requirements are not well defined and attempts to gather requirements from customers and end users have had limited success. Also, prototype software development has benefits in helping customers and end users gain confidence in the developers' proposed implementation. In the latter instance, seeing is believing.

As noted earlier, a prototype does not have to be throw-away. Modern software development methods, tools, platforms, and architectures make it possible to build enhancements incrementally across any number of development iterations. Several Agile methods, such as Scrum, SPIRAL, and dynamic systems development model (DSDM), are based on the notion of adding functionality to existing software products over multiple, short-term iterations.

It is important to go into a prototyping effort with the realization that it might be a complete failure. The organization has to be willing to let it go, take the chapters learned, and move on.

Process Overview

Developing a prototype is, in effect, a miniature version of a normal software development cycle (see Figure 5). Prototyping typically includes the following steps:

1. **Gather requirements**
   Can include business process, functional, user and systems requirements

2. **Develop systems design and architecture**
   At this stage, the developers initially assess high-level architectural requirements, components, and interfaces that support the needs of the business and end users, without worrying about implementation details. As the architecture solidifies, providing the list of necessary software or system components and their interrelationships, the developers turn their attention to creating the design document and design specifications describing the purpose, functionality, and interfaces required in each specified component, and methods for construction.

3. **Develop prototype**
   The prototype is developed and should at a minimum include mockup screens with data and user input fields displayed. RAD tools and modern integrated development environments (IDEs) allow developers to create more complex prototypes that demonstrate application logic and data flow. At the highest end of functionality, the organization may obtain the rights to use a COTS application for a trial period.

   DID YOU KNOW?
   RAD tools implement GUI-based user interface (UI) development and modeling tools to speed development and de-emphasize the need for coding. An IDE is a suite of tools that have a standardized GUI to help software developers create, build and test their code.

4. **Obtain customer feedback**
   Customers, end users and perhaps other stakeholders need to commit their time to evaluate the usability of the prototype and provide feedback on enhancements that lead to a better functional product.

5. **Update requirements and design specifications**
   The development team updates the requirements documents and design specifications from the input gathered in the user and customer review sessions. If there is a significant change in requirements, it may be necessary to resubmit a proposal to expand the scope, budget, resource and time commitments originally approved. If the customers and end users are not satisfied with the results of the prototype, then it may be necessary to repeat steps #2 through #4 and come up with a satisfactory result, before the sponsors authorize the start of development.

6. **Initiate development cycle**
   Assuming the executive sponsors and customers agree on the updated requirements and adequate funding and resources are available, the

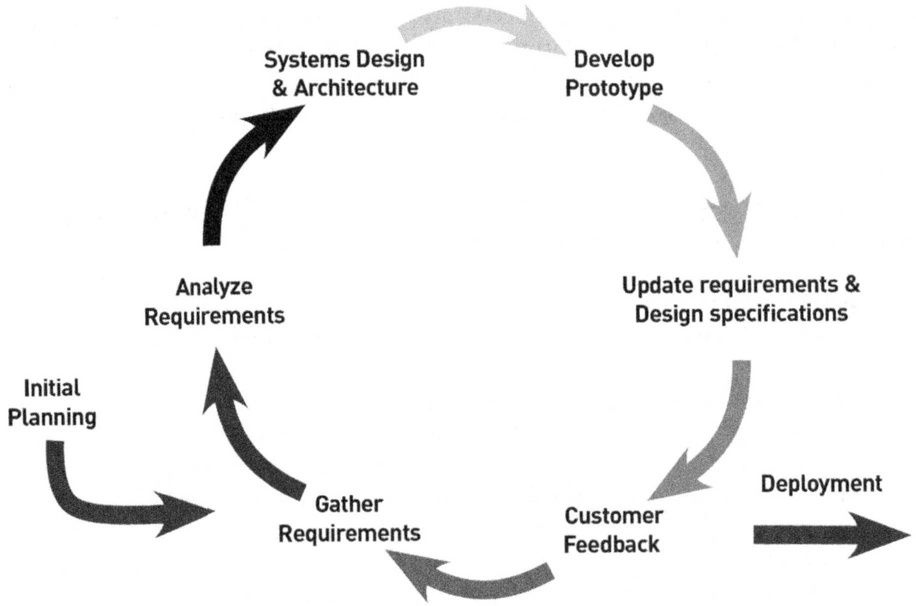

Figure 4. Prototyping Process

Usability Engineering
In the field of Usability Engineering – a discipline that works on improving human-computer interfaces, prototypes are used to improve software or system usability from an end user's perspective. Usability has two forms:

1. Improving the user interface (GUI)

2. Simulating enhancements that provide users with desired and useful capabilities (e.g., helping end users and customers do things that are difficult, if not impossible, with the methods and tools they have today.)

Types of Prototypes
In general, there are two types of prototypes, throwaway and evolutionary. A throwaway prototype, as the name implies, is not meant to become a software product released into production or as

a commercial offering. It is simply used as a simulation to get customer and end-user feedback and to test the practicality of design concepts.

In contrast, an evolutionary prototype is an extension of the IID concepts where each prototype is incrementally adding value to a final product. This concept gets a working version of the software into end users' hands sooner than traditional waterfall-based development methodologies.

Solution Applications

Prototypes are not meant to be, in of themselves, deployable solutions or applications. Their primary purpose is to quickly and economically evaluate design concepts before full-scale development. They are also used to provide customers and end users with a mock-up of the proposed solution to obtain their feedback before committing to full-scale development.

Evolutionary prototypes are typically created as the initial effort within an iterative development life cycle. However, the distinction between throwaway and evolutionary prototypes does not need to be made at the start of a prototyping effort. Initial prototyping efforts that are shown to be ineffective can be thrown away and started over before committing too much time and resources to a particular design.

On the other hand, a surviving prototype can evolve into a full-scale iterative development cycle, with features and functions added incrementally until the final solution is realized and ready for release into production. In other words, early versions of an evolving prototype may not include all the capabilities customers ultimately desire; however, if a design is successful, the prototype can evolve incrementally to the point that it serves as a baseline for a future funded development activity and release.

Pros and Cons of Prototyping

Table 4 shows the pros and cons of using a prototyping approach to development.

Table 4

Pros	Cons
Reduced risk of getting it wrong	Not an outright replacement for requirements analysis
Reduced costs and time spent going down the wrong path	Prototypes are not intended to be a complete system
Better quality in the end product	Could mislead customers and sponsors on the complexity and costs associated with building a complete solution
Customer-focused approach and better visioning of the final solution	Prototypes, especially throwaway prototypes, are sunk costs that may not lead to a return on investment (ROI)

Prototyping Summary

In conclusion, prototyping as a practice is encouraged when there is uncertainty in customer and end-user requirements or the potential design for a system. Executive sponsors must be willing to view their investments in prototyping as sunk costs, since the initial design may prove to be impractical or not in alignment with customer needs.

There are also perceived, if not real, time delays and costs associated with this practice when compared to traditional waterfall and Agile practices. On the other hand, prototyping is a practical risk mitigation strategy that helps prevent catastrophic failures that may be associated with premature development, before requirements are well understood and validated by end users and customers.

From a usability perspective, prototypes can support development and testing of user interface enhancements. Prototypes help the development team test the feasibility of proposed workflows and data flows within an application to support a new or changed business process. And, prototypes are used to assess alternative implementations of functional enhancements proposed by end-users.

Prototyping is always a subprocess within other SDLCs, including modern waterfall, hybrid and Agile-based development approaches. Examples of Agile SDLCs that use prototyping concepts are the DSDM, SPIRAL, RAD, and RUP.

IID MODEL

Much like Agile, the IID model is more of a concept than a fully specified SDLC. The IID approach has evolved independently, has been around since at least the mid-1950s, and is included as part of most modern development approaches. This section highlights only some of the more important concepts in this development approach.

Before I start, let's do a quick definition of terms:

The term *iterative* implies that the development of a product is broken up into smaller and relatively frequent development cycles. Each iteration may implement new functionality or enhance some part of the existing code. Each development iteration can include one or more traditional SDLC processes. Client delivery, deployment, and provisioning activities can be included in a development iteration, but more often are separated as distinct product release activities.

The term *incremental* implies an approach to build different parts of a software system at different times and rates and then integrate the components as they become available to form a complete product offering. In other words, each integration enhances the value of the software or system with new capabilities, features, and functions.

In a modern context, the terms *iterative* and *incremental* are combined to describe a development pattern where software is gradually enriched with new functionality and improved through a recurring pattern of development and release cycles. Also implied with the IID pattern is continuing customer and end user feedback to guide the direction of development in terms of needs and overall usability.

Two Development Strategies in One

The terms incremental and iterative are often put together to describe a single methodology that employs iterative development and feedback loops but blurs the distinction between activities devoted to functional improvements versus those that improve the quality of software. However, it is important to note that early advocates of the IID approach, such as Alistair Cockburn, saw iterative and incremental as representing two distinctly different types of development strategies.[11]

Specifically, Cockburn viewed incremental development as a "staging and scheduling strategy" where components of the overall software solution are built at different times and rates and then incrementally assembled when completed. In contrast, he viewed iterative development as a "rework scheduling strategy" where time is made available to review and improve the system. [12]

In software engineering, the term *code refactoring* is used to describe an iterative process of rewriting existing code to improve its structure, readability, and extensibility without changing the external behaviors of the original code. Iterative development used to refactor code makes the existing code better, with the goal to improve nonfunctional attributes. Nonfunctional attributes are quality measurements based on operational and performance criteria. Therefore, iterative development can be viewed as a software or system quality improvement activity.

In contrast, incremental development improves product functionality by breaking up development into parts, and then integrating them to form a complete solution. Each component is built separately, tested, and iteratively improved until it's ready for release. Unlike the traditional waterfall model, where all components are built and released at once, each incremental build can be deployed as a separate release. This approach is referred to as an incremental build model.

Both IID activities are necessary and can go on in parallel. However, Cockburn argues that the term IID has blurred the distinction between these two development practices. He believes it's important to maintain the distinction when directing work on a project to improve both the quality and the functionality of a product or system under development.

Covering the Basics of IID

The basic concept of an IID approach is a rapid and customer-centric evolution of a product. Rather than try to determine all the requirements and the overarching system design at the start of the project, the development team instead progressively exposes new requirements and design needs through a set of iterative prototyping and development stages that incrementally add to the solution's capabilities.

The IID model divides the development of a complete solution into logical increments based on developing components that implement desired functionality. Cross-functional teams (representatives from different organizations who have different skills, rolls, and responsibilities) work the development of each component from gathering initial requirements all the way through deployment. The goal is to ensure that each team has the right resources to build the right product. For example:

- Business executives and product owners provide an understanding of business requirements and priorities
- Domain experts & end Users bring an understanding of business processes and needs
- Business analysts bring methods and tools to gather and document requirements
- Software and systems architects bring knowledge and skills to determine the components and interfaces necessary to build a working solution around the requirements
- Software developers bring the skills to design, build and test a solution
- Database designers bring the skills to design, build and test the application's database
- Cybersecurity experts bring skills to ensure that software and networks are properly secure from hacking and other attacks
- Software testers bring skills to ensure that the developed products conform to requirements and design specifications and meet all desired performance criteria
- CM specialists ensure that the artifacts produced during the development iterations are properly versioned, managed, and protected to safeguard the organization's IT assets and ensure long term supportability and sustainability of the products
- IT operational staff who must install, configure, support, and sustain the system once it is deployed into production

IID Unified Process

A unified process represents one of the better-known methodologies to support IID implementations. UP is an architecture-centric and use-case driven approach to IID. UP is considered an architecture-centric approach due to the focus on incrementally developing the solution as a series of functional and integrated components. Use cases are a form of requirements analysis that visually and textually describes a series of actions (a.k.a. events steps) taken between an actor and a system that is designed to support a goal of the action. The actor can be a person or another system of some type.

As shown in Figure 6, the unified process includes four iterative and incremental phases: inception, elaboration, construction, and transition.

The inception phase is used to identify the desired scope of work, functional and non-functional requirements, and potential risks.

The elaboration phase has a focus on developing an architecture that supports the requirements and mitigates identified risks. The components identified in the architecture are incrementally developed and integrated within the construction phase. Construction iterations typically include customer and end user reviews at the end of each cycle to solicit their feedback.

Finally, an incremental solution is deployed to the production environment during the transition phase, and the IID process starts once again.

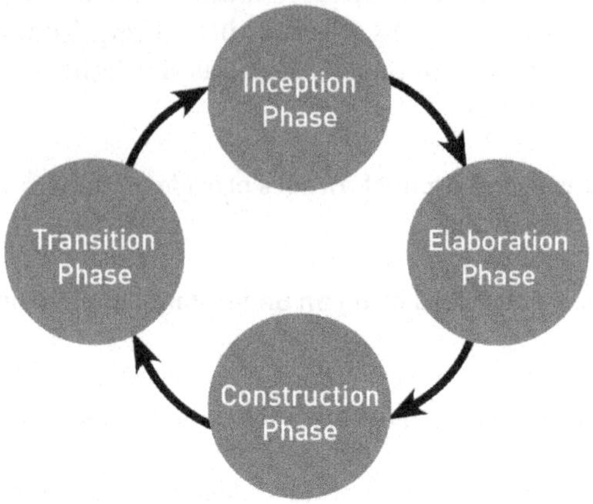

Figure 5. Unified Process Phases

The work associated within each IID phase can be divided one or more iterations, depending on scope and complexity. All iterations are compressed, or "timeboxed," into relatively short and equal durations. This timeboxed approach evens out the work among the various participants to keep the flow of work moving evenly across the phases while eliminating the potential for work to pile up and queue.

The IID phases are cyclical, but also operate in parallel. In other words, business analysts can be working on scoping, requirements analysis, and risk identification activities for an upcoming release while architects are working on the architecture and designs for the next development cycle. In the meantime, developers are working on coding and testing activities for the next release, while the transition team is moving a previous release into production. These iterative development cycles allow the product or system to evolve incrementally over time, with ongoing customer guidance influencing new enhancements and modifications across each iteration.

IID Release Approaches

There are two forms of IID models from a release perspective:

1. **Delayed release**—A comprehensive software release is developed over multiple development iterations without attempting to make each iterative release deployable

2. **Potentially shippable release**—each development iteration is designed to produce a potentially shippable product (PSP)

3. Potentially shippable product—text here?

A delayed release does not attempt to create a functionally complete product and holds back deployment until all the iterations are sufficiently complete to satisfy a larger set of customer and end-user requirements for deployment. In this scenario, each iteration is, in effect, a limited prototype on the way to delivering the final solution for deployment.

With a potentially shippable release, each development iteration delivers a functionally complete product. However, even in this scenario, the IT organization may hold off releasing the product into production for logistical reasons. Modern Agile approaches incorporate variations of these two IID models.

In practice, most IID implementations evolve and test products across each iteration but hold off on deployment until fully implementing a specific set of customer-desired features. Delaying product releases is necessary for many if not most businesses, particularly on large enterprise applications that support cross-departmental business processes, as the organization cannot support new product deployments at the same pace as the development team can deliver new functionality.

For example, end user training on product enhancements and business process changes may be required to support a new product release. The organization may need to limit deployments on a trial basis to ensure that the deployment doesn't introduce large-scale problems that are costly and difficult to resolve. Communication is also an important component of a successful rollout of a new or enhanced product that supports cross-organization business processes.

In some organizations, such as online retail stores, there may be a need to roll out new software releases quite frequently. In my book, SDLC Modern Practices, I introduce continuous integration, testing, delivery, and deployment capabilities that enable releases over very short intervals, even on the order of several times a day. These capabilities integrate and automate SDLC processes.

IID Process Overview

With an IID model there is typically an initial planning phase to get the development team together and capture high-level business, customer, and end user requirements, and to create a backlog of desired product enhancements desired for the release. The architects determine the components and interfaces necessary to achieve the functional requirements identified by the business analysts. The development team works through their planned development iterations to produce and test the code that implements the desired functionality. The transition team oversees production and delivery of administration and support documentation, training aids and training services, plus configuration and delivery of the system into the production environments.

Collectively, the activities in a complete IID cycle mimic those found in the traditional waterfall-based models, but the work is broken up into smaller chunks, worked in parallel, and managed across smaller timeboxed iterations. Note that the development team allocates time within each construction phase iteration for software testing (including unit, integration, QA, user acceptance, performance, and production testing, at a minimum).

Table 5. Historical development of the PDSA cycle

Who	When	What
W. Edwards Deming	1950-1990s	An early collaborator with Walter Shewhart evolved the Shewhart Cycle as a 3-step model into a 4-step iterative management method known as the plan-do-check-act (PDCA) cycle, and later modified into the plan-do-study-act (PDSA) model. Both models are used to facilitate c ontinuous and incremental process improvements. (a.k.a. Deming Wheel).[8]
T. Gilb and R. Zultner	1976 & 1988	Discussed the use of PDSA concepts to software development in their respective books.[9]
Craig Larman and Victor R. Basili	2003	A definitive study on the history of IID. Shows how IID concepts have their roots in the Shewhart Cycle and PDCA/PDSA cycles.

Customers need time to evaluate and comment on the state of the product at the end of each development iteration. Their feedback can include requests for enhancements to the existing product, or the identification of completely new requirements as the customers and end users visualize new ways to use the product. This input goes to the development team and the product owner and informs their views on future development priorities.

In practice, the product backlog for the release is broken up into 2 to 4-week development cycles to complete the entire release. Subject to limitations on budgets, resources and deployment time frames, there can be any number of these development iterations allotted to a release. Products are released into production as each incremental offering has sufficient functionality to satisfy customer expectations while minimizing the impact of supporting a new release.

IID is built on PDCA/PDSA Concepts

The iterative and incremental development concepts build upon the concepts of the plan-do-check–act (PDCA) and plan-do-study-act (PDSA) cycles that were initially defined as approaches to business process management and improvement. Though the origins of the PDCA/ PDSA models precede modern software and systems development practice, the concepts are still relevant. Table 5 provides a quick historical introduction to the development of the PDCA/ PDSA cycle concepts.

The developers of these cycles, Walter Shewhart and W. Edwards Deming, established a disciplined, four stage approach to address inadequate business processes. The first step is to investigate the problems and then define a strategy to fix a broken process or current business need. Once the strategy is defined, the next stage is to implement whatever changes are necessary to execute the strategy. In the third stage, the results are checked to compare the results of the change against pre-defined performance metrics to see if the changes are achieving the desired objectives. The primary difference between the PDCA and PDSA models is that the later includes the analysis activities to see what was learned from the change that can help further refine the business process. In other words, did the changes achieve the anticipated results? If not, why not? And, have we obtained new insights from the change activities? Finally, in the fourth stage, if the changes meet the performance criteria, the organization acts to document and implement the new practices. The PDSA cycles starts again, based on the analysis and findings of the Study stage.

Table 4 - Historical development of the PDSA cycle, provides a historical view on the development of the plan–do–check–act (PDCA) and plan-do-study-act (PDSA) cycles. The cyclical and four stage improvement process is found in most process improvement approaches, in project management, and in iterative development strategies.

Timeboxed Iterations

In the IID models, the term *timeboxed* is often used to describe a relatively short development time frame, usually between 2 to 4 weeks for each iterative development cycle. There is no strict rule on the amount of time in which a development cycle occurs, but in modern practice, individual development cycles focus on the implementation of some set of functionalities that fits within the desired timeboxed duration.

The primary objective is to ensure that customers and users have a chance to view and perhaps use the software before making too many updates that would be difficult and expensive to change. The customers, having seen the software, are then able to add to or further refine their requirements based on the insights they gain from seeing and working with the preproduction release of the software.

The Importance of Customer Feedback

One of the insights I discovered early in my IT consulting and project management career is that users very often can't see the total vision for a new software offering, or how the software can affect their work, until they get their hands on it in one form or another. When a new release comes out under a traditional waterfall-based approach, the resulting conversation between customers and developers can go something like this:

> *Software user*: "Wow! I do like these pull-down menus, and graphics, and icons... and these tabs are perfect for breaking up our tasks by business function. And, yes, the data fields and workflows seem to be working as we requested. Nice! Hmmm, but can the software also have a menu option to create a new type of product offering that marketing is releasing, and perhaps a new tab to relate the order information to cost, inventory, and product fulfillment information... etc., etc., etc.?"
>
> *Software developer*: "Yeah, the software could do all those things... if only you had told me that at the beginning of the project, back when we were in the requirements phase."

This is a perfect example of where the waterfall method has limitations. Users do not understand what is possible, or even what is needed until they have something to see or use in a real-world context. In the waterfall approach, assuming new enhancement requests are funded, the users may have to wait another year before those enhancements can be made available to them in a new product release.

Evolving Applications

IID approaches work best where there is uncertainty in requirements and design, or requirements are subject to change, and the project sponsors are willing to provide time, and access to customers and end users to assist in reviews during the iterative development process. The involvement of customers and end users is the critical success factor for this approach to software development. On the positive side, their input leads to a better, more customer-focused solution.

Under IID, executive sponsors and customers must have some tolerance for risk, and a strong belief that the product can evolve to meet their requirements. They still must dictate budgets, resources, and schedule constraints. On the positive side, they will see results more quickly. Instead of attempting to define all the requirements up front, they must let go and allow the team to expose the requirements iteratively and understand that the product may likely evolve in directions not originally anticipated.

When organizations are willing and able to allow their IT-based systems to evolve and mature incrementally, the development team can embark on a continuous development process. In such cases, new releases of large-scale enterprise applications can be made available on an ongoing basis, across one fiscal year to another. High-level requirements consistent with corporate strategies still must be defined each fiscal year, along with approved budgets and resource allocations. However, the development team may see little change otherwise to mark the move from one program or project approval to the next.

In many applications, this is a better scenario than waiting for yearly releases for new product enhancements. Also, under IID, IT development projects for large enterprise applications can become a fixed cost. As long as the executive sponsors and customers are happy with the results, the iterative development activities can go on for the entire life cycle of the project, with new releases coming out every few weeks or months, depending on customer preferences.

Today, there are several formalized IID methods to choose from, such as the DSDM, XP, Scrum, DAD, and Scaled Agile Framework (SAFe). These methods not only offer a chance to provide midcourse corrections, they also enable a solution to evolve as requirements are better understood.

When employing IID processes, executive sponsors and customers must have a mindset that values incremental improvements over shorter periods—as opposed to expecting a home run and getting everything they want, the first time at bat. The "home run" strategy has more risks.

Pros and Cons of IID

Table 6 - Pros and Cons of iterative and incremental development approaches, provides a list of the pros and cons associated with using an IID approach.

Table 6. Pros and Cons of iterative and incremental development approaches

Pros	Cons
Customer feedback and course corrections are more frequent than with waterfall.	Focus on iterative development in smaller increments may make it difficult to see the big picture
Iterative builds make it easier to find and resolve functional and design issues before they become entrenched in the larger system.	Focus on short-term improvements at the expense of detailed requirements and systems analysis can lead to poor design.
It is less difficult to change project scope or product requirements in the middle of the overall life cycle	IID requires customers and end users to stay engaged, and provide feedback, in periodic product reviews between each iterative development cycle
In some cases, functionality is deployed incrementally.	It is costly and time-consuming to deploy the final solution.
Incremental and iterative processes reduce risk by finding and fixing problems as the project proceeds, as opposed to facing a catastrophic failure at the end.	Executive sponsors and customers do not know what capabilities, features, and functions they are getting until the product is released.

IID Section Summary

In conclusion, the IID model is a customer-focused approach to building software products and systems. Another benefit is customers and end users can obtain access to new features and functions in shorter time periods with an incremental release schedule. However, nothing comes for free. Customers and end users must be willing and able to commit time to review and provide feedback after every iterative release.

Moreover, as new requirements and enhancement requests drive each development iteration, executive sponsors and customers do not know what capabilities, features, and functions they are getting for their money until the final release. IID approaches are a fundamental principle behind Agile-based practices found within the major Agile-based development methods, such as the DSDM, RAD, RUP, SAFe, Scrum, and SPIRAL.

RAD MODEL

As the name implies, the term *rapid application development* is a strategy that emphasizes development activities over initial planning and requirements gathering activities. As software development tools evolved, this approach became a practical alternative to traditional waterfall-based development methods.

RAD is largely a tools-based prototyping strategy. GUI development tools are often used to expedite development of mock-up screens that customers and end users can use to review and provide their input at the earliest stages of the system design. Also, reusable software modules enable rapid implementation of working data fields and data flows to test out more complex business and user concepts.

With RAD, planning, requirements analysis, and design activities are not as structured as with waterfall and often run concurrently with development. In this respect, RAD is very much a trial and error approach that incorporates iterative prototyping. It is also a great approach when the end-user interface and overall usability are key design factors that need to be tested by end users.

Modern programming languages, especially those with predefined and reusable software modules and GUI builders, have made RAD a practical approach to enable rapid prototyping. RAD is most appropriate when requirements are changing and not well-defined, and there is little need to have strict control over the final capabilities or design of the product. RAD is not typically considered a practical solution for very large projects where design, architecture, integration issues, and budget and schedule constraints dictate more control over the project.

Covering the Basics

RAD approaches radically depart from waterfall approaches by combining requirements analysis and design with development activities. Planning too is minimized, as the trial and error method of rapid prototyping would quickly undo any previously defined plans.

In a RAD model, more time is spent on developing prototypes than on initial planning and design. Some practitioners choose not to develop detailed requirements and design documents, preferring to spend their time and resources on developing the software incrementally, directly from customer and user inputs, allowing them to capture an evolving set of requirements in the software itself. In short, RAD advocates the embrace and support of a change in requirements that comes from having customers and end users directly involved in iterative prototyping and development increments.

Proponents of RAD would also claim that the very act of developing software provides new insights on how a product can better support a business process or the work of end users. In their view, rapid prototyping gets to these insights more quickly and more assuredly than with traditional requirements gathering sessions.

Supports BPR and Continuous Process Improvements

RAD techniques and development tools became available at about the same time as Business Process Reengineering (BPR) concepts became popular. BPR practitioners would note that software applications should not just automate the current business practices, as a bad process is still likely a bad process when automated. On the other hand, RAD's trial and error approach to rapid prototyping enable new insights and approaches for business process improvements (BPIs) to be tested and implemented that otherwise might not be considered due to the time and expense associated with other development methods.

Early RAD proponents also recognized that it is better to find problems early when it is easier and less expensive to fix them. Early discovery and fixing issues just make sense. Under traditional methods, the software would not be tested as a system or by users until after completion of the final build of the software.

As software development tools evolved, especially those with predefined and reusable software components and GUI design tools, rapid prototyping became mainstream, particularly on software development projects in support of BPR and BPI efforts. With RAD methods and tools, developers and business or domain experts could collaborate and try out different business process and information access concepts with minimal time and cost.

A Tool-based Agile Strategy

By and large, RAD is a tools-based approach to developing software. RAD tools provide a common graphical interface and menu-driven environment for development, as opposed to the older command line types of coding environments.

OO programming tools were early adopters of RAD concepts, supporting the need to define, create and manage reusable software components. Other development platforms, such as visual development tools, business process, data modeling tools, and GUI development tools, are also included in the RAD approach. Integrated Development Environment (IDE) and Web Integrated Development Environments (Web IDE) are modern terms for integrated suites of tools that provide RAD-type capabilities.

RAD tools enable developers to quickly build screens and menus to demonstrate and allow users to work out various information display, data flow, and workflow concepts. RAD prototypes are built incrementally, and the development team only focuses on a narrow set of requirements for each iteration, with each prototype incrementally adding to the overall software solution.

Moreover, the prototypes created by the RAD tools are not necessarily throw-away products. RAD prototypes can serve as a baseline for development, once the team is given the go-ahead to start developing the final software solution. In this sense, RAD tools are a foundation capability for both prototyping and IID approaches.

RAD-based prototypes are built in timeboxed phases, delivering new functionality incrementally, over shorter time frames. As a result, customers receive new software enhancements more frequently, and with a faster evolution of business supporting capabilities.

Process Overview

RAD as an SDLC has evolved with advances in technology. Historically, the term is claimed by D. Dinadasa and Dan Gielan, who were with the New York Telephone Company's Systems Development Center in the mid-1970s. In their vernacular, the term RAD was a general term used to describe prototyping and iterative development approaches where insights gained during the development process serve as feedback to requirements in time to impact the current application development cycle. In this context, RAD was one of the early development concepts to be promoted as an alternative to traditional, top-down development approaches.

Later, in 1991, James Martin used the term RAD to describe a formalized process of four phases that included Requirements Planning, User Design, Construction, and Cutover (See Table 7). [13] The user design and construction phases implement prototyping and iterative development practices. However, Martin's approach also employs the use of JAD practices and CASE tools to create descriptive models of requirements to help expedite development of the final applications.

Table 7: Rad Model Tools	
Requirements	The gathering and analysis of business, customer and end-user information and performance support requirements
Graphical User Interface (GUI) Designs	GUI Builders employ graphical control elements that can be rapidly assembled to create application user interfaces
Data Models	An analysis of required data objects, their relationships, and flow of information
Object Models	An analysis of how the app should handle information received from external sources
Process Models	An analysis, description and often a graphical depiction of current and future state processes

These characteristics of RAD, rapid prototyping, and IID, are preserved today; however, what's changed are the capabilities of the development suites and tools that have evolved to support RAD approaches. For example, modern tools provide capabilities to record requirements, develop the GUI, model data and process flows, and implement object-oriented development (OOD) tools, which support the development of reusable software components. These tools all feed the iterative development life cycle until the product is ready for release.

In this graphical model, I have shown a set of requirements gathering and modeling activities, as inputs to an iterative development cycle. These activities include:

- **Capture business and user requirements**
  - The gathering and analysis of business, customer and end user information and performance support requirements
- **GUI designs**
  - GUI builders employ graphical control elements that can be rapidly assembled to create application user interfaces
- **Data models**
  - Analysis of required data objects, their relationships, and the flow of information
- **Object models**
  - Analysis of how the app should handle information received from external sources
- **Process models**
  - An analysis, description, and often a graphical depiction of current and future state processes

In James Martin's original diagram, shown on the right side of Figure 7, the output of requirements planning activities serves as an input to an iterative cycle of user design and construction activities. The circle around these two sets of activities represents an iterative development process that includes application development, testing, customer reviews, and feedback. Through each iteration, the inputs are revisited, based on the feedback obtained from customers and end users. As the software matures and is ready for release, it exits the iterative development process into the Production Turnover phase. In other words, the software or system is released into production.

Solution Applications

RAD is ideal for small to medium-sized projects where the requirements are not well understood or subject to change, and management has the patience and risk tolerance to allow for a bit of uncertainty in the final deliverables of the project. The solution should be of a size and complexity that fits within a relatively short time span of perhaps two to three months. The implementation of short development cycles in RAD is mostly an issue of budgeting and risks. The longer the release cycle, the more costs that are involved, and there is more risk for the paying customer who will fund a deliverable item that does not fit within their strategic requirements.

As with all IID processes, there can be multiple development iterations in a 2 to 3-month release cycle, with each iteration incrementally adding value to the software or system. Since developers can build prototypes quickly, RAD is useful on projects where customers are trying to define new or enhanced business processes and system support capabilities. The prototypes can be used to try out different concepts for information access, data flows, workflows, and system designs. Most importantly, customers and end users must make the time to be involved throughout the development process. They must proactively guide development in support of their requirements.

Pros and Cons of RAD

Table 8 provides a list of the pros and cons typically associated with the RAD Model.

Table 8. Pros and Cons of the RAD Model

Pros	Cons
User feedback incrementally and rapidly drives the development of desired features, functions and capabilities.	Customers/end users must devote their time to software reviews, or this methodology absolutely won't work.
With customers and end users in control, there is reduced risk of getting it wrong.	The approach is so different from waterfall that the change scares both customers and IT. executives
RAD is more likely to create a better overall product within time and budget constraints.	Management has less control over the outcome, which can be viewed both as a positive and negative, depending on the application and viewpoint.
Well suited for small to medium-sized software applications where user interface and usability are the primary business drivers for the application	Focus on user interface over system design can have negative consequences, affecting the overall software architecture.
There is less chance of having a catastrophic failure with frequent reviews and go/no-go decisions.	RAD is not a good choice for large-scale enterprise applications where business needs and related control over requirements and design—to achieve strategic enterprise mission, goals, and objectives—are the critical business drivers for the software investment.

RAD Section Summary

RAD is a tools-based approach to development. The tools employ GUI-based user interface (UI) development and modeling tools to facilitate rapid development. RAD methods and tools are particularly useful in development projects that incorporate prototyping and IID strategies.

As a rapid prototyping and incremental development approach, this trial and error development concept is in stark contrast to the highly structured and sequential waterfall SDLC. Where waterfall attempts to codify and enforce a specific set of requirements that support a desired or unchanging business process, RAD methods and tools allow the understanding of requirements and business processes to evolve. This evolution is driven by feedback from customers and end users as they interact with the software prototypes. This type of interaction with the evolving software also enables a more informed understanding of what is possible, including the testing of concepts that might not otherwise be considered.

Throughout this chapter, I have noted that customers and end users very often need to see and use the proposed software before they can properly envision how the system can be employed to support their usability and business processes requirements. The RAD approach certainly supports this human behavior. RAD approaches are often enhanced with the use of modern development suites and tools, supporting Agile concepts with rapid prototyping and IID capabilities.

BIG BANG MODEL

This development methodology is mentioned frequently as a potential development strategy; however, there is almost nothing written in the way of historical background. This methodology seems to have appeared just as suddenly, and from nothing, just like the cosmological version for which it is named. (See Figure 6)

On a more serious note, the Big Bang approach has no defined process. Instead, the focus is on encouraging creativity by not implementing rigid processes, organizational structures, or bureaucracy. Resources are assigned to work on a project with, at best, a high-level objective. Projects using this approach are typically very small, with one or two developers, or maybe a small team working in a "skunk works" type environment—an innovative undertaking by a small group of people outside the normal research and development channels within an organization. They are responsible for any requirements analysis, project planning, coding, and testing. Since they are developers, they spend the bulk of their time and attention on coding the application.

This type of approach is useful in four scenarios: 1) prototyping; 2) academic learning practices; 3) a skunk works type development activity, and 4) one or two guys working out of a garage to create the next killer app. The point is to place all the developers' focus on creativity.

Covering the Basics

If we went by the name of this software development life cycle alone, Big Bang, we might think this is some robust, large-scale methodology for software development projects that includes everything but the kitchen sink. However, nothing could be further from the truth. Rather, the name serves as an analogy to the cosmological model of the big bang, that much like our universe, something great can come from seemingly nothing. Like the cosmological Big Bang, as shown in Figure 8 - Big Bang model, great and wonderful things sometimes come from very small beginnings.

The fact is, the Big Bang model lacks formal structure and is usually reserved for small development projects with the goal of building a quick prototype to show concepts under consideration, or to demonstrate a student's ability to develop functional code that successfully implements specific capabilities that were assigned to them by an instructor.

A startup company with limited resources might also take this unstructured approach as they build their initial product offerings. In a larger company, this approach works best as skunk works operations where highly skilled developers are given time and resources, but not much oversight or structure, to see if they can come up with the next new thing. The reality is, there are many more failures than successes with this approach. Then again, it might only take one major success to make such an effort worthwhile.

If a useful product results from a Big Bang development effort, it must be removed from the skunk works environment and matured into a releasable product under a completer and more formalized SDLC approach. There is too much opportunity for the product to have bugs or miss important usability and functionality as a commercial offering.

Process Overview

With the Big Bang model, there is little formal planning, no formal requirements analysis, and no detailed project plan. The developers are given a general goal and simply set to work developing the software code.

This model is typically used by very small development teams—again, perhaps only one or two developers, who work without formal oversight. With such a small team, and with only high-level objectives defined, there is little need for formal planning and scheduling. Since there are limited resources for requirements gathering and analysis, the developers spend the majority of their time on software design and coding activities. Requirements analysis is limited to whatever time the developers can commit to it. There is not even formal testing, so the risk of failure is great.

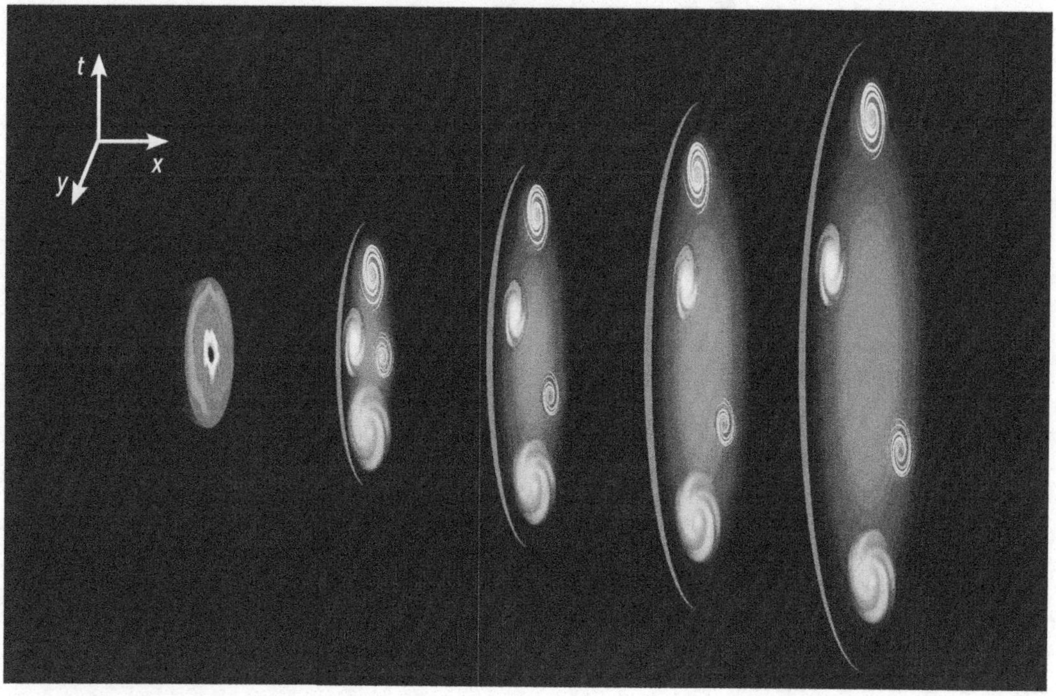

Figure 6. Big Bang model

Solution Applications

So why in the world does this model exist? This model is not even a model. There are no best practices, processes, or procedures to follow. It is essentially developing from scratch on a blank canvas. However, as it turns out, there are genuine applications for this model.

One of the biggest challenges software development teams face is to extract a decent set of requirements from the potential customers and end users of a software product. Until the customers have something they can see, it is often hard for them to envision what features, functions, workflow, or information they require from a software application.

One of the practical uses of the Big Bang model, in a business setting, is to develop a quick prototype as a throwaway effort to test out various concepts. It's a form of prototyping without any structured approach. Such a prototype can give end users and customers something to view and use, and ultimately start the right level of conversations to flesh out the initial requirements for a formal software development effort. Also, as I mentioned previously, such an approach might be used in a skunk works type operation where skilled developers can experiment with the goal to invent something new and useful.

The other typical application is in an academic setting, where time and resources are limited. This model offers a practical approach for readers to practice developing code to solve a proposed problem or scenario.

Of course, another practical use for the Big Bang model is by a developer hoping to be the next Steve Jobs or Bill Gates, working out of his or her garage to build the next killer app.

Pros and Cons of the Big Bang Model

Table 9 provides a list of the pros and cons associated with the Big Bang model.

Table 9. Pros and Cons of the Big Bang Model	
Pros	Cons
Simple	High risk of failure
The quickest route to development	Few, if any, requirements to guide the development team
No management overhead	With only high-level objectives and no detailed requirements guiding development, not likely to result in a useable product without additional user feedback and modifications
Minimal costs and resources	High probability of throwing the effort away
Promotes creativity over imposed structure and oversight	Doesn't provide management and oversight
Can be used to demonstrate software development skills in an academic setting	Not a good model for large and complex software development projects with rigid requirements

Big Bang Section Summary
The Big Bang development model is not a formal SDLC. Rather, it is a descriptive term for small development activities, and development teams with limited resources, that have the goal to create big things from small beginnings. The emphasis is on enabling creativity with minimal structures or oversight. The four primary uses of a Big Bang project are building limited prototypes, using a skunk works environment to build the next big thing, demonstrating skills in an academic environment, or initiating the start of a small business or commercial application with very limited resources.

AGILE MODELS

Introduction
Agile is a philosophy that serves as an umbrella for a large number of software development life cycle approaches advocated as alternatives to the traditional waterfall and other so-called "heavyweight" software development approaches. The values and principles of Agile were developed and promoted by a group of software developers in 2001.

Covering the Basics
Agile is at the other end of the spectrum from waterfall. Where waterfall involves a single, highly structured development cycle that can extend out over a year or more, Agile software development techniques deliver incremental changes over smaller time intervals, with much more customer involvement in iterative requirements, design, development, and test cycles.

Agile software development methodologies were, and in some cases still are, seen by many as a radical departure from traditional waterfall approaches, which had dominated the software development landscape for so many years. As pointed out in this section, Agile practitioners have a different philosophical approach to software development that contrasts greatly with traditional waterfall approaches. Where waterfall is focused on documentation, structure, and control to achieve predefined and unchanging outcomes, Agile practices have a focus on accommodating evolving customer/user needs and embracing change.

Process Overview
The 4 core values and 12 guiding principles of the Agile approach to creating software are copied in accordance with site permission as follows.[14/15]

MANIFESTO FOR AGILE SOFTWARE DE VELOPMENT

From the Agile Manifesto, the values promoted under Agile are stated as follows:

"We are uncovering better ways of developing software by doing it and helping others do it. Through this work we have come to value:

- **Individuals and interactions over processes and tools**
- **Working software over comprehensive documentation**
- **Customer collaboration over contract negotiation**
- **Responding to change over following a plan**

That is, while there is value in the items on the right, we value the items on the left more." Table 10 lists the Agile Manifesto Authors:

Table 10. List of the Agile Manifesto Authors		
Kent Beck	James Grenning	Robert C. Martin
Mike Beedle	Jim Highsmith	Steve Mellor
Arie van Bennekum	Andrew Hunt	Ken Schwaber
Alistair Cockburn	Ron Jeffries	Jeff Sutherland
Ward Cunningham	Jon Kern	Dave Thomas
Martin Fowler	Brian Marick	

PRINCIPLES BEHIND THE AGILE MANIFESTO

Besides the four values quoted on the previous page, the Agile Manifesto cites the following 12 Principles that further refine Agile practices:

- **Our highest priority is to satisfy the customer through early and continuous delivery of valuable software.**
- **Welcome changing requirements, even late in development. Agile processes harness change for the customer's competitive advantage.**
- **Deliver working software frequently, from a couple of weeks to a couple of months, with a preference to the shorter timescale.**
- **Businesspeople and developers must work together daily throughout the project.**
- **Build projects around motivated individuals. Give them the environment and support they need and trust them to get the job done.**
- **The most efficient and effective method of conveying information to and within a development team is a face-to-face conversation.**
- **Working software is the primary measure of progress.**
- **Agile processes promote sustainable development. The sponsors, developers, and users should be able to maintain a constant pace indefinitely.**
- **Continuous attention to technical excellence and good design enhances agility.**
- **Simplicity--the art of maximizing the amount of work not done--is essential.**
- **The best architectures, requirements, and designs emerge from self-organizing teams.**
- **At regular intervals, the team reflects on how to become more effective, then tunes and adjusts its behavior accordingly.**

Solution Applications

Agile is a concept where developers value direct customer input and welcome changing/evolving requirements—by embracing IID methods and eschewing excessive structure and documentation in favor of flexibility and adaptability. Where waterfall approaches attempt to create the entire solution in one development cycle, Agile practitioners prefer to break up a development project into bite-sized chunks, allowing the solution to evolve over several customer-directed iterations.

Agile is best suited for projects where the requirements are not well-defined or subject to change, and the executive sponsors have some tolerance for uncertainty in solution requirements, or even schedules and budgets. The best practice is to fund Agile-based development projects as a fixed-cost activity, with continuing releases available every few months for the life of the product.

Pros and Cons of Agile SDLC Methods

Table 11 lists some of the pros and cons associated with Agile development practices.

Table 11. Pros and cons of Agile development practices

Pros	Cons
Agile empowers developers, customers, users, and other stakeholders to work together to develop the best solutions.	Initial architectures and designs may not be optimal as new requirements emerge.
A customer-centric approach results in better outcomes for end users.	Without defining the requirements up front, scope creep can set in, causing budget and schedule overruns on projects that have specific goals under budget and schedule constraints.
A flexible approach can accommodate changes in requirements.	The sponsors paying the bills may have little tolerance to fund projects with so much uncertainty of the outcomes.
Frequent releases allow deployment of new capabilities in shorter development cycles, and more often.	Customers and end users may find it difficult to make the time to review and provide feedback on multiple iterations through the development life cycle.

INTERESTING FACTS:

Scrum is one of the better known and more widely used Agile software development methodologies and can trace its history back very well to the continuing efforts of two men, Jeff Sutherland and Ken Schwaber.[16] Like many other software development methods, the methodology has been continually adapted and modified by individual practitioners. As an example, with a simple search on Amazon.com, interested readers can find more than 2000 results returned on the subject of Scrum.

As a result of this outside influence, one can find any number of variations on the basic model. In this overview, in chapter 8, Development Phase, we use the Scrum methods as an example of how to implement Agile methods within a modern corporate SDLC.

Modern Agile

As with all the other systems development concepts presented in this chapter, Agile continues to evolve. Numerous writers have published books to explain their views on methodological approaches to implementing Agile values and principles. There is more than one certifying body for Scrum. The Project Management Institute (PMI) collaborated with the Agile Alliance® to define an Agile Practice Guide to integrate Agile development approaches within the traditional project management model. Moreover, numerous IDE and development tools vendors have produced extensions to their products to support Agile development practices.

However, a growing body of Agile practitioners has come to believe Agile has gone the route of the traditional waterfall-based practices, becoming bloated with unnecessary processes, practices, and ceremonies that distract from the simplicity and efficiencies of Agile. They also believe the values and principles of the Agile Manifesto were due for an update.

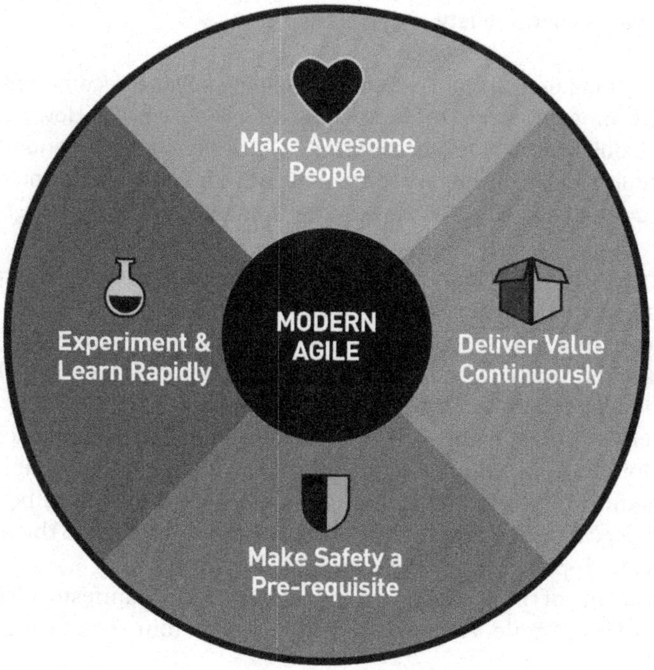

Figure 7. Modern Agile Principles

As a result, a community of users has developed a set of new principles that they believe provides a "simpler, sturdier, [and] more streamlined way to be Agile."[17] Under modern Agile development approaches, the focus shifts to "producing exceptional outcomes and growing an outstanding culture," and the original four values and twelve principles, as outlined in the Agile Manifesto, are replaced with the four guiding principles shown in Figure 7 – Modern Agile Principles, and further described below:

- **Make people awesome**
  - "In modern Agile, we ask how we can make people in our ecosystem awesome. This includes the people who use, make, buy, sell or fund our products or services." We learn their context and pain points, what holds them back and what they aspire to achieve. How can we make them awesome?"
- **Make Safety a prerequisite**
  - "Safety is both a basic human need and a key to unlocking high performance. We actively make safety a prerequisite by establishing safety before engaging in any hazardous work. We protect people's time, information, reputation, money, health, and relationships. Moreover, we endeavor to make our collaborations, products, and services resilient and safe."
- **Experiment & learn rapidly**
  - "We learn rapidly by experimenting frequently. We make our experiments 'safe to fail' so we are not afraid to conduct more experiments. When we get stuck or aren't learning enough, we take it as a sign that we need to learn more by running more experiments."

- **Deliver value continuously**
    - "Anything that isn't delivered isn't helping anyone become more awesome or safe. In modern Agile, we ask ourselves, 'How could valuable work be delivered faster?' Delivering value continuously requires us to divide larger amounts of value into smaller pieces that may be delivered safely now rather than later."

Modern Agile borrows from the lean methodology, which I will discuss later in this chapter and again in my book titled *Frameworks for Agility*. The modern Agile approach is considered a lightweight process in contrast to the processes developed in support of the original *Agile* Manifesto's values and principles. Indeed, Modern Agile is not a process at all; it is a principle-based approach to make organizations Agile and responsive.

There are no defined roles, responsibilities, charts, dictated ceremonies, or professional certifications in Modern Agile. Modern Agile does not provide guidance at all on how work is performed, what skills are needed, IT policies or standards that should be followed, or what tools the organization can use to develop software. It's simply a "back to the basics" strategy for developing nimble, high-performing, and awesome organizations.

Alistair Cockburn, one of the cosigners on the original Agile Manifesto, also believes it's time to get back to the basics of Agile. His new model focuses on four core "words.

- Collaborate
- Deliver
- Reflect
- Improve

Again, the idea is to lose all the baggage of heavy-weight processes and instead focus on critical success criteria that drive Agile organizations. I don't disagree with any of these concepts. Processes, activities, methods, tools are all enablers. However, they can also get in the way if the burden and bureaucracy get in the way collaboration, teambuilding, safe work environments, and efficiency of efforts.

I applaud the efforts of the Modern Agile Community to get back to streamline development activities and return to the use of lightweight and customizable approaches to development, based on the individual needs of each project. I also applaud the principles outlined under the Modern Agile. They are people centric.

When I started in my career as an IT consultant and in sales, my mentor, Jim O'Malley, would constantly state: "People do business with people." There is not a greater truth in business, and I have never forgotten his admonishments to stay focused on this singular principle. The principles of Modern Agile would fit Jim O'Malley's views on business quite well.

I suspect that in another 15 years or so, maybe less, any number of authors, consultants, and vendors will likewise come up with concepts, methods, and tools to support the principles of Modern Agile. I believe it is human nature to improve on things. However, sooner or later, these methodologies will become heavyweight and bloated.

So, I'm guessing my readers have to be asking, why is this author writing an entire series of books on systems development if he agrees that the values and principles of the original Agile Manifesto and the newer Modern Agile principles are relevant? The answer is relatively simple. I'm helping readers understand the total scope of work involved in systems development. On any given project or development iteration, a team may choose a different set of methods and tools to support the specific needs of the customer and their project.

Much like we may have a selection of clothing in our closet from which we choose what to wear on any given day, the project team should have a sufficient selection of methods, tools, and engineering practices they can select from. In my opinion, this is part of project tailoring. Moreover, to complete the analogy, an organization's mature SDLC framework is the closet, and an organized closet, at that.

In my research on this book, the biggest takeaway I got was that there are numerous methods, engineering practices, and tools to assist in the development of software and complex IT systems; the trick is to borrow the right bits and pieces from multiple approaches and blend them to meet the specific needs on any given project or development cycle. Don't attempt to implement a "one size fits all" strategy. Moreover, above all, keep your projects simple and aligned with the values and principles outlined in The Agile Manifesto or the principles of Modern Agile.

Agile Section Summary

Agile is a set of values and principles, and not a specific or standardized SDLC process or method. Over the years, numerous methods and tools evolved to support Agile values and principles. Some of the Agile-based methodologies have become heavy-weight. If your corporate policies, development processes, and ceremonies become overly burdensome and inefficient, you've probably lost your way. Go back and revisit the values and principles of Agile.

While Agile does not promote specific methods or tools, the basic concepts of prototyping, IID, RAD, and even the Big Bang development practices, still serve as foundational development concepts. Modern IDEs support developer team collaboration, prototyping, and IID concepts often associated with Agile practices.

CHAPTER SUMMARY

This chapter started with several descriptions for the acronym SDLC and then provided an overview on the evolution of software development foundational concepts. Understanding the foundational concepts of software development makes it easier to evaluate alternative approaches, as most methodologies build from these foundations. The six foundational approaches include.

- Traditional, sequential (waterfall)
- Prototyping
- IID
- RAD
- Big Bang
- Agile

The foundations discussion started with an introduction to the traditional plan-drive and linear sequential approach of the Waterfall model. Prototyping are low cost, low risk and potentially throw away development projects or assessments of commercial-off-the-shelf (COTS) solutions.

Prototypes help reduce risks in both the traditional ad agile approaches to software development.

The iterative and incremental development (IID) concepts are foundational to most agile-based methodologies. Iterative development is a strategy to break up development into numerous ad time-boxed cycles to reduce development complexities, improve efficiencies, reduce the number of intractable bugs, and obtain frequent input from customers and end-users to guide further development needs and priorities. The term incremental describes the concept that each development iteration offers new increments of useful functionality.

Rapid application development (RAD) describes the use of tools, in particular computer-aided systems engineering (CASE) tools, to reduce cycles times and complexities in software development and requirements analysis. An objective of early CASE tools was to deploy a number of modeling tools built upon a central repository as an integrated system to build and maintain software. The modeling tools and central repositories managed information related to project management, architecture and design specifications, object models, code specifications, test cases and test results, data diagrams and reports. In a modern context, CASE tools have given way to integrated development environments (IDE) which provide a suite of software development tools built around a common graphical user interface (GUI) and supporting one or more programming languages.

The Big Bang approach to software development is used in prototyping and educational environments, where the goal is to focus on developing something new and interesting while reducing prescriptive rules and processes or formalities that would impede the developers' work.

Finally, at a conceptual level, agile was introduced as the set of values and principles outlined in the Agile Manifesto. Agile is not a process, and its not a set of tools. Rather, Agile is a philosophy that stresses the need to reduce the complexities and inefficiencies of the traditional plan-driven and linear-sequential development models. Agile puts the focus on improving customer and team relationships and interactions and on building products that work. Advocates of agile deemphasize the importance of tools, or following prescribed processes, or creating documentation, as the means to building better products. Instead, Agile seeks closer alignment with customers, ensuring the teams stays focused on meeting existing priorities and being responsive to the client's changing priorities and emergent needs.

03. SDLC Process Improvement Approaches

This chapter offers an introduction to selected SDLC process improvement, collaboration, and automation methods that enhance or support the Agile Model. This chapter briefly introduces model SDLC practices that help streamline, integrate and automate SDLC processes. This chapter serves as a precursor to the material that is presented in the fourth book in this series, on SDLC Mature Practices.

CHAPTER OBJECTIVES

After reading this chapter, you should be able to describe each of the following modern SDLC practices:

- Six Sigma,
- Kaizen,
- Kanban
- Just in Time (JIT),
- Lean,
- Scrumban,
- DevOps,
- Continuous Integration
- Continuous Testing
- Continuous Delivery & Deployment

The processes are listed alphabetically, not historically.

AUTHOR'S INTRODUCTIONS

Chapter 2 introduced foundational SDLC concepts that are commonly applied across most if not all the SDLC methodologies you may encounter. Now I am going to introduce development concepts that are not foundational—in the sense that they are absolutely required, but rather augmenting the development approaches within the SDLC method.

Most of the development concepts introduced in this chapter did not come from the software industry. Many have their roots in the field of statistical process control and psychology and were initially implemented in manufacturing industries to reduce complexity, waste, and inefficiencies. Many of the techniques introduced in this chapter have found a renaissance of sorts in the software industry, such as KAIZEN, JIT, Kanban, and lean manufacturing concepts.

All the concepts introduced in this chapter facilitate Agile-based SDLC methodologies, helping an IT organization implement practices that are in alignment with the values and principles outlined in the Agile Manifesto. So, while the core SDLC concepts presented in Chapter 2 are foundational, we can think of the development practices outlined in this chapter as augmenting SDLC practices.

AGILE-FRIENDLY PRACTICES

Any automation method that supports the Agile model can enhance an Agile development approach and, as appropriate to the needs of the organization, should be described and maintained within a corporate SDLC as a formalized and approved development process. Again, my intent is not to recommend or advocate any of these approaches, but rather to provide introductory information and point to source reference materials for those who want a deeper dive on any of these Agile-friendly SDLC approaches.

All the practices mentioned in this section, plus several others, are revisited in more detail in my *DevOps* book. The books listed as reference source materials are all in my library. Plus, I have read the articles listed; I wouldn't recommend reading something I haven't looked at myself.

Test and Behavior Driven Development

> **Test-driven development (TDD)** supports agile development practices by breaking down development tasks into very small cycles built around individual requirements. Moreover, before writing any code, the developers create a test script that cannot pass in integration testing without a new module that successfully fulfills the requirement. Once the test script is written, and the integration test fails, the developer writes the section of code that allows the test to pass.
>
> This strategy accomplishes two goals. First, it ensures there is not exiting code that would already cause a false-positive pass. Second, it ensures code is written correctly from the start to fulfill the underlying requirements.
>
> **Behavior-driven development (BDD)** is an Agile software development process that encourages collaboration between development team members, users and customers. A hallmark of the process is the use of a domain-specific language (DSL) to facilitate communications and understanding between technical and non-technical participants. e.g., A DSL uses English-like sentences to describe product requirements.
>
> User stories are a common example of a DSL. A user story expresses needs from the point of view of a user, explaining what they want to be able to do, and why. Ideally, each user story documents one succinct requirement. A user story follows this format:

"As a <*type of user*>, **I want** <*describe desired capability*>, **so that** <*reason*>."

BDD supports TDD by documenting customer and user requirements in a concise and mutually understandable format. The DSL format also simplifies the process of translating individual requirements into associated test scripts and test results, all in one brief statement.

DevOps (Development-Operations)

The term DevOps is a clipped compound of the two words: *development* and *operations*. The original context was to formalize the need for organizations to facilitate communications between IT development and operations-oriented organizations. The goal was and is to streamline delivery of new products into production and, in reverse, facilitate the exchange of information back to the development teams to make sure the right products are being built and to improve product supportability and sustainability. In the past, these were traditionally siloed organizations. A conceptual model of a mature DevOps environment, encompassing all Build, Release and Operational processes, is shown in Figure 8.

DID YOU KNOW?

The term *DevOps* is credited to Patrick Debois, a software development consultant who named a conference in 2009 DevOps Days. The concept of DevOps was then popularized in a book titled *The Phoenix Project* in 2013.

The early focus on collaborations and communications between development and operations organizations remains a critical component of DevOps. However, real improvements come from the synching of SDLC processes all the way from concept to end-to-life of a product. Process integration and automation are key components of a modern process improvement activity. IT-based process improvements are no different than other BPI initiatives in that regard.

However, integrated and automated processes by themselves do not guarantee efficiency or effectiveness in terms of achieving corporate strategies and goals. The real improvements from DevOps come from the elimination of wasted, non-value-added efforts—as is the objective of Lean methodologies, while also reducing WIP that only adds clutter and complexity—the objective of JIT systems.

By analogy, think of how a manufacturing assembly line works. Everything is meant to work in lock-synch, with all products flowing from one position to the next at the same rate. If one area of the line slows down, the entire assembly line has to slow down with it.

Do you remember the old *I Love Lucy* episode where Lucy and Ethel find a job wrapping chocolates at the candy factory? At first, they keep up, but then the line speeds up, and they can't stay up with it. To keep up, they start eating some of the chocolates—which is rather funny in its absurdity, but that still doesn't help, and absolute chaos ensues. To complete our analogy, Lucy and Ethel are the limiting constraints in their dramatized candy manufacturing line.

While this sounds silly and we laugh at the antics of Lucy and Ethyl, a real-life broken process can be equally chaotic and dramatic. For example, I started my consulting career helping manufacturing companies improve their processes. I was at a toilet paper manufacturing facility one day when part of the assembly line failed. If you can imagine how much toilet paper is used every day, then you can probably imagine how fast the output is in these types of manufacturing facilities. Until someone was finally able to get to the panic button to stop the assembly line, toilet paper rolls went flying everywhere at very high rates of speeds. It made Lucille Ball's chocolate factory scene seem tame by comparison.

It makes no sense to integrate and automate IT processes across the life cycle of a product, if the processes are inefficient and out of synch with each other or have a high probability of failing. In other words, we must constrain the flow of work across SDLC processes, or any business process for that matter, to ensure that the most efficient development operations occur at a sustainable rate. Over time, the organization should continue to prioritize and make incremental improvements on those processes where bottlenecks in the work occur.

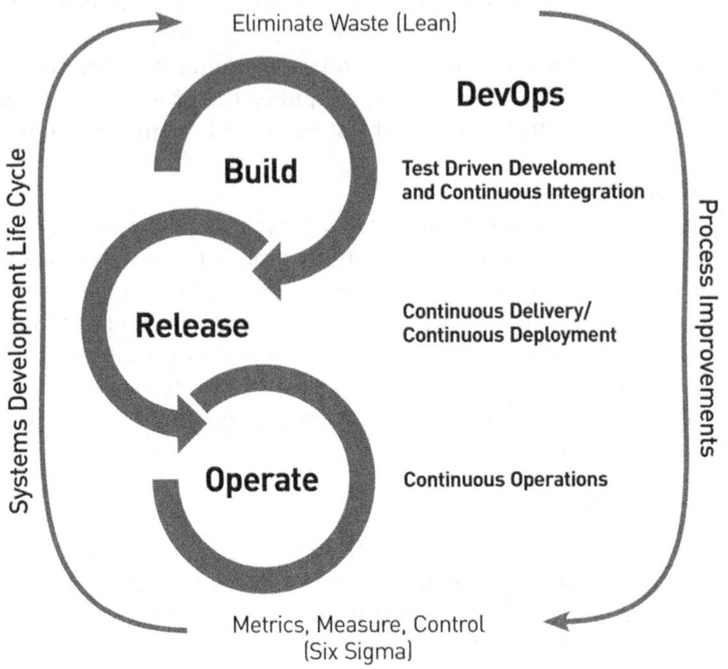

Figure 8. Conceptual model of a mature DevOps environment

It should be clear that SDLC processes are just as important as ever when implementing DevOps capabilities. It's impossible to implement DevOps without a sound understanding of the organization's SDLC processes. How can we integrate and automate processes we don't understand?

To recap, a mature DevOps environment facilitates communications and collaboration across participating organizations and has efficient, integrated, and highly automated IT processes spanning the entire SDLC. In such environments, it's not unusual to see daily or more frequent software product releases into production. This development approach is presented in much greater detail in my book on *DevOps*.

Theory of Constraints

The concept of making incremental process improvements by reducing bottlenecks is well established. Though not the originator, Eliyahu M. Goldratt popularized these concepts as the "theory of constraints" in his 1984 book, *The Goal*. *The Goal* was written as a fictional story to demonstrate how the theory of constraints can work in a manufacturing environment. In 2013, Gene Kim, Kevin Behr, and George Spafford released a book, The Phoenix Project: A Novel about IT, *DevOps, and Helping Your Business Win,* that rewrites the theory of constraints as a fictional story based on an IT-centered scenario.

For those who are interested in exploring this subject further, while Goldratt is credited with coining the term "*theory of constraints*," his concepts were built from previous works by others. For example, in 1958, D.E. Broadbent, an experimental psychologist from the United Kingdom, published a book titled *Perception and Communication* in which he noted that individuals have a limited amount of attentional resources they can use at one time.[18]

Humans, in effect, filter what they hear, and then decide what they'll pay attention to, comprehend, and retain. This process limits the amount of information that can be absorbed and retained at any given time. Broadbent used as an analogy a bottle filled with sand turned upside down to help explain how the filtering process restricts the flow of information that can be processed. The flow of sand out of the bottle is constricted by the bottles' neck, hence the term *bottleneck*.

Broadbent's theories applied to the limitation's humans have in processing information at any given time. However, the initial application of the concepts of bottlenecks in the physical world is credited to Wolfgang Mewes in his 1963 paper, titled "Power-oriented management theory," and another paper titled 'Energo-*Kybernetic* Cybernetic System (EKS), 1971"—later renamed "Engpasskonzentrierte Strategies."[19]

Since I don't read German, I asked a dear friend, Dale Carpenter, whose wife and extended family are from Germany, help me interpret Wolfgang Mewes' title and the intent of the article. Dale noted that the literal interpretation of the title is "concentrated" or "enriched bottleneck." However, his son Mark noted that he would translate the title as "bottleneck-focused." Mark further notes: "In this case, the word "*konzentriert*" is utilized to mean "focused." Moreover, the "Engpasskonzentrierte *Strategie*" [article] is addressing [use of] power and dependency ratios instead of cost and profit ratios.

Continuous Integration (CI)

The term *continuous integration* has been around since 1991 when Grady Booch, then with Rational Software (a company later acquired by IBM), introduced his Booch Method for OOD.[20] CI is also one of the 12 core principles of XP.[21]

Software code under development is maintained in a source code repository. When developers need to work on a new requirement, they pull a copy of the source code (a.k.a. code base) from the repository and begin their work. As they make changes, the developers will test their new code as an independent unit, and then as an integrated part of the larger code base.

In its most generalized form, CI is a practice that involves frequent merging of all development code back into the trunk of the code base (a.k.a. main branch, baseline, mainline, and master code) within the source code repository. This code integration process can occur daily or with every single code change made in a modern CI environment.

Software code is developed incrementally to fulfill the requirements and design specifications. Before starting to code on a new piece of software, or to extend an existing piece of software code, the developer needs to review the requirements to understand what the code needs to do in the form of a logical operation or procedure. They also must understand what data is required as an input to the code, and what the expected results should be when the code is executed.

Best practice is to write a test initially, called a unit test, that will validate the code's functionality before they start writing the necessary software code or making changes to existing code. The code is then written, compiled (turning the code into a machine-readable format), and then unit tested to ensure that it functions properly.

The unit test is also a small fragment of software code that provides the inputs to the code being tested and returns the results after execution to help the developer validate the success of their code. In effect, the unit test stands in for other software components that may not be developed yet.

Then, the code needs to be integrated with the mainline source code to ensure that the new code doesn't break existing functionality in the larger code base. This process is referred to as a software build. Build automation is a set of scripts that help facilitate the build process. The process of creating new software builds iteratively, as new code changes are made, is referred to as continuous integration.

A software build usually involves compiling multiple source code files and linking other object files in the correct order. Some files that have been previously compiled and not changed or affected by the new build may not need to be recompiled.

If one developer holds a copy of the source code for any length of time, other developers may have, in the meantime, updated the code base. For this reason, a developer needs to make sure they have the most current code base before integrating their code. As the software expands, the build process can become complex, tedious, and time-consuming to update the code base. Modern tools have evolved to support and automate the build process, making continuous integration a less burdensome process.

DevOps, CI, and CD are not possible without a mature versioning and CM system. A development team with parallel development activities releases new software builds asynchronously over time. In other words, components put into the source code repository do not have to follow a sequential release pattern. Individual parts of the code base may be checked out any number of times, reworked, and checked back in. While one part of the code is being worked on by one set of developers, other components are worked on by other developers. Every time a component is merged with the mainline code, a new product release is formed. This type of complexity leads to a situation that has become known as "integration Hell" within the software industry.[22] Under continuous integration processes, if a release is scheduled to be tested or put into a deployment environment, it's critical that we know exactly what components go with it. Version control methods ensure proper versioning of all components developed over time, while CM identifies all the components and their specific versions that make up the release.

Figure 9 offers a simplified visual example of product versioning. The figure is a simplified model, as there can be any number of branches, components, and subcomponents coming off the main trunk line. A modern CI tool automatically monitors the source code repository and sees new software updates as they come in. The CI tool uses input from the version control and CM tools to understand the new/revised source code's place in the code base, and then run the automated build and test scripts to compile, test, and link the related artifacts to create the new build.

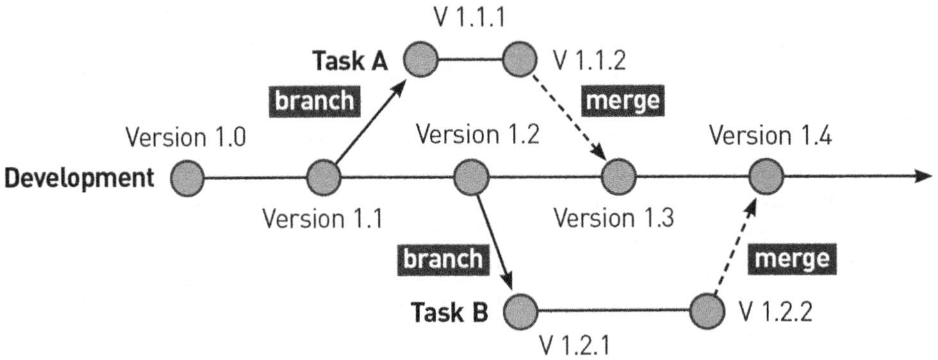

Figure 9. Version control as part of CI

Continuous Testing (CT)

Continuous testing is a part of the software delivery pipeline capabilities implemented by organizations that are integrated and automating their SDLC processes. The goal of CT is to obtain immediate feedback on any bugs or defects at all stages of the development process, and across all development iterations. In the next section, continuous delivery concepts are discussed, CI makes it possible to integrate and automate testing by standing up virtual servers and testing software and test scripts on demand.

In my book titled *Frameworks for Agility*, I spend quite a bit of time discussing the types of testing necessary to put a product into production safely. For those who are interested, on the *Software Testing Material* website, the author, S.M. Rajkumar, maintains a list of more than 100 potential types of software testing on his website.[23]

The majority of the testing approaches cited on the *Software Testing Material* website have unique applications and are not required on most projects. Based on my personal experience and observations, the following testing types are most commonly required in on large scale commercial and federal software development projects.:

- **Component testing**
- **Functional testing**
- **Integration testing**
- **Regression testing**

- **Load/performance testing**
- **508 compliance testing**
- **QA testing**
- **Security testing**
- **System end-to-end testing**
- **User acceptance testing**

If you go on to read my book title SDLC Mature Practices, you will see how continuous delivery capabilities allow an organization to stage development and testing environments on demand, when and as needed. In effect, CD implements an automated provisioning workflow on top of the development and testing processes. Some organizations are automating provisioning and testing, which is referred to as *continuous testing*.

Continuous Delivery & Deployment (CD/D)

Both continuous delivery and continuous deployment are extensions to the CI concepts and capabilities. Where modern CI tools integrate and automate the traditional SDLC processes from, potentially, requirements gathering through coding and testing, CD methods and tools integrate and automate the provisioning of computing and software assets and deployment into staging and production environments.

In this context, think of CI as practices and tools for automating front-end SDLC processes while CD has a focus on practices and tools for integrating and automating back-end SDLC processes. Another way to think about it: CI facilitates development activities, while CD adds capabilities to facilitate operational activities. So, it should be clear from these contextual views that CI and CD together form the basis for creating a mature DevOps environment.

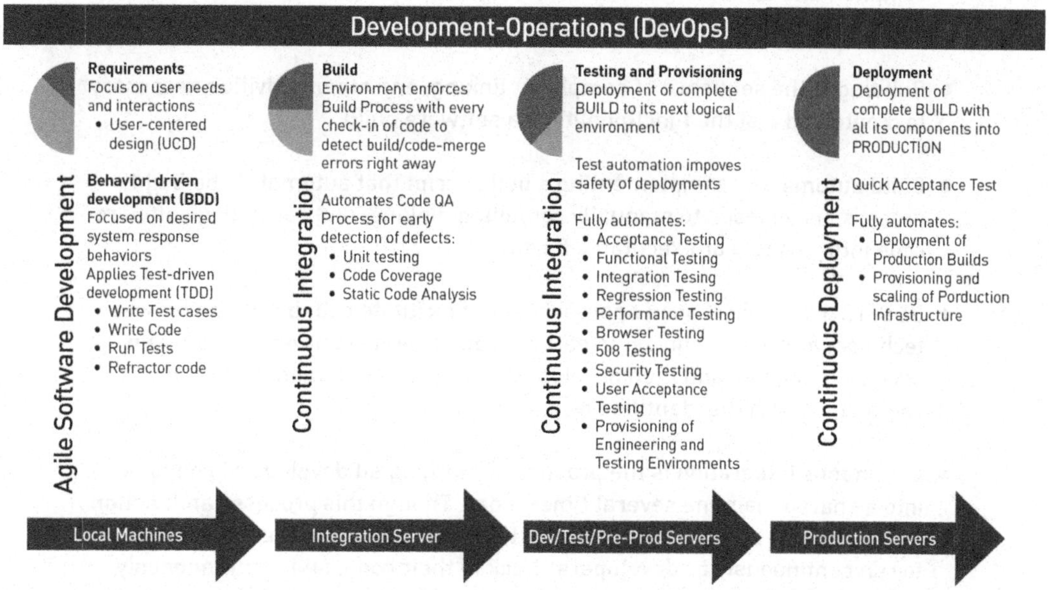

Figure 10. DevOps Model with SDLC Automation

The primary difference between continuous delivery and continuous deployment is the act of human intervention before moving a new software or system release into production. A development organization that implements a continuous delivery strategy places a human intervention step before allowing a new release to be moved into a production environment, while automating the release into the production environments is referred to as continuous deployment.

There are several reasons for maintaining human interventions. They provide an opportunity for the development organization to perform a final quality check before delivery of a new product or enhancement to a customer. The executive sponsors may require stage reviews as the product develops and final approvals before release. The organization may prefer to limit deployments to a select community of users, and then slowly expand the release as functionality and performance are proven within across the user base. Finally, the customer may not be able to accept continuous releases. For example, if the product is an enterprise application supporting critical business processes, the customer may need to sync delivery with formal process releases and training of end users and product support personnel.

On the other hand, there are situations where continuous deployment makes sense. One of the biggest drivers for continuous deployment is to improve or maintain a competitive advantage on an ongoing basis, such as with online retail sales or online service delivery companies. Even in those situations, it's not unusual as a risk mitigation strategy to see an initial limited deployment, to a select community of target market customers, before implementation of a full-scale release.

Recap of DevOps and SDLC Automation Concepts

DevOps is a concept that promotes communications and collaborations between development and operations-oriented IT teams. But DevOps also implies an integration and streamlining of SDLC processes to achieve greater efficiencies. Figure 12 provides a comprehensive view of a DevOps environment that implements SDLC automation and integration concepts. Let's do a quick review of what we've learned so far.

- Compiling is a process that turns human-readable source code into machine-readable object code and saves it as an object file.
- Linking is the process of taking one or more object files, generated by the compiler, and combining them into a single executable file, a library file, or another object file.
- Building is the sequence of compiling, linking, and other activities necessary to create and test the functionality of a software build.
- Build automation is implementing a build script that automates the build process, as opposed to manually compiling, linking, and packaging software code into a usable or executable form.
- BDD/TDD are agile-based approaches that facilitate collaborations between technical and non-technical to gather requirements in a common domain specific language, and then develop test scripts that demonstrate code compliance with the identified requirements.
- Continuous integration is the practice of merging all developers' changes into a shared mainline several times a day. Though this process can function as a manual task, in a modern context, CI tools automate the build process to run continuously as developers check in their code, and simultaneously run the appropriate unit tests to ensure that the code base still functions properly as it is rebuilt.
- Packaging is the process of structuring and combining executable files and documentation related to a software product or system to facilitate deployment of the application.

- Automated provisioning eliminates the manual tasks necessary to scale the infrastructure (hardware, networks, and software) and deploy software services to a large and distributed set of end users. Automated provisioning capabilities are most often employed as a cloud-based computing service and removes the key bottlenecks in production deployments.

- Continuous testing automates the provisioning of testing environments on demand, and execution of relevant testing types, to immediately identify defects and bugs at all stages of the development process, and across all development iterations. Test results are returned immediately and automatically to the developers for review. This process both expedites and simplifies the bug/defect resolution process.

- Continuous delivery automates build, testing, and provisioning processes to deliver the artifacts related to the release of a software product or system into a staging area that mimics the intended production environment. Manual intervention is necessary to move the artifacts into the production environment. This is the baseline set of capabilities for a mature DevOps environment.

- Continuous deployment has all the features of continuous delivery but removes the manual intervention step required to move the artifacts into the product environment. In other words, every software release that passes the automated tests is immediately and automatically deployed into production.

ADOPTED LEAN DEVELOPMENT STRATEGIES

Lean development strategies largely originated to support the manufacturing industry, initially in Japan and at the Toyota Motor Company, before their strategies were adopted internationally within the manufacturing industry to maintain a competitive advantage. These concepts are introduced here to lay the foundation for explaining how the concepts are now applied within the software development industry.

The concepts *lean*, *JIT*, *Kanban*, and *kaizen* originally came out of manufacturing industries, the principles and techniques behind them have found their way into software development in several Agile-based software development methodologies. However, the term "lean software development" was initially introduced in a book by the same name, written by Mary and Tom Poppendieck, and published in 2003.[24]

LEAN

The term *lean* was coined in the late 1980s by Jim Womack, PhD., then at MIT's International Motor Vehicle Program to describe Toyota's business.[25] Jim Womack remains involved with developing lean concepts as a cofounder and Senior Advisor to the Lean Enterprise Institute.[26]

The central idea behind lean is to "maximize customer value while minimizing waste," and "create more value for customers with fewer resources."[27/28] In their book, *Lean Thinking*, James Womack and Daniel Jones note that an organization embarking on a lean transformation effort must start with an understanding of the three fundamental business issues of purpose, process, and people.[29] The lean process includes five steps:

1. Identify value,
2. Map the value stream,
3. Create flow,
4. Establish pull, and
5. Seek protection.

The goal of lean is to improve quality and reduce waste. More to the point, lean processes take a holistic view to add value to the customer—regarding producing with quality the right set of products and the elimination of wasted activities of any form that do not add value to the customer.

JUST IN TIME (JIT)

JIT is considered an inventory management system. The ideal situation is to sync delivery of materials, goods, and labor to arrive or be replenished exactly when needed in a production process. Most important, JIT provides a method to constrain and sync production operations and deliveries from suppliers with customer demand.

The goal of the JIT approach can be summarized in one succinct statement: "The complete elimination of non-value-added activities and waste."[30] The distinction between the focus of lean and JIT processes is subtle, but important. Lean process improvements seek to improve customer value across the organization and its value chain, while JIT processes improve production efficiencies and elimination of waste (e.g., excessive materials and WIP, plus maximal but not over-utilization of resources) in sync with customer demand.

From this description, it should be somewhat obvious why JIT is sometimes considered a subset of lean concepts. Where lean seeks to use process improvements to improve customer value across the value chain of an organization, JIT is more limited in reducing excessive inventories and improving efficiencies in production.

Of course, we live in a dynamic world, and the objectives of JIT present a constantly moving target. Therefore, continuous improvement and proactive response to change go together with any attempts to maintain the benefits of JIT.

The initial JIT concepts were developed in Japan by Toyota and were manifested in the Toyota Production System (TPS). Toyota still maintains information on TPS on their website and note that TPS is also referred to as a lean manufacturing system or a JIT system.[31] The two primary concepts of JIT include:

- **Quality must be built in during the manufacturing process** [32]
- **Making only what is needed, when it is needed, and in the amount needed** [33]

Poor quality creates waste in the form of lost time, resources, and money spent on lost product and rework. Another form of waste is the excessive inventory of materials and work in process—which causes inefficiencies in the development environment, undue complexity, and the potential for developing products that are no longer relevant or useable.

KANBAN

Kanban is a system (a.k.a. "supermarket method" based on the originating roots of the technique) that Toyota implemented to control the flow of work. In the simplest of terms, a Kanban card provides a pull-based system that syncs the flow of work to demand. The Kanban card serves as a signal to a preceding process, flowing in reverse from the new order request, indicating parts that have been used and need to be replenished.

I will discuss the implementation of JIT and Kanban concepts in software development in the Scrumban section. For now, understand that a Kanban system limits the work in process across a development facility and constrains the flow of work to demand and the speed of the slowest participating process. As mentioned in the previous discussion of the theory of constraints, such constraints counterintuitively improve productivity and efficiencies by eliminating waste.

KAIZEN

Kaizen is another concept that emerged from Japan and Toyota manufacturing, and is translated into English as change (kai) good (zen) but is more loosely translated as "improvement." A kaizen approach to business is one of continuous improvements. This is the modern use of the term kaizen at Toyota.[34/35]

Lean Software Development

Lean practices preceded the development and release of the Agile values and principles outlined in the Agile Manifesto. Jeff Southerland, one of the co-founders of Scrum, is on record noting that lean concepts "tremendously affected the evolution of scrum." [36] Other modern SDLC frameworks, such as the SAFe® incorporate lean concepts.

Moreover, as noted earlier, Mary and Tom Poppendieck originated the term lean software development in a book with the same title that was released in 2003. In their book, the Poppendieck's articulated 7 Principles of lean software development, which include:

- Eliminate waste
- Build quality In
- Create knowledge
- Defer commitment
- Deliver fast
- Respect people
- See/optimize the whole

In addition to the 7 principles, the Poppendieck's introduce 22 tools of lean to aid and tailor the implementation of lean practices in software development.

Scrumban

Scrumban is a hybrid model developed by Cory Ladas that integrates the concepts of Scrum with Kanban's JIT development strategies.[37 / 38] Ladas acknowledges his proposed Scrumban methodology builds upon Lean principles, as well as the underlying Scrum-based Agile methodology.

In his book, Ladas asks a fundamental question: "Under what condition is lean software development possible?" In response, he proposes two axioms:

> **Axiom 1:** It is possible to divide the work into small value-adding increments that can be independently scheduled.[39]
>
> **Axiom 2:** It is possible to develop any value-adding increment in a continuous flow from the requirement to development.[40]

Ladas' book includes a series of essays that detail his thoughts on best practices to implement Agile and lean concepts in software development. My biggest takeaway in reading his essays is he presents ostensibly sound strategies for implementing Kanban techniques via task cards or tokens on a Scrum-oriented task board. The critical focus of his methodology is using the Scrumban system to constrain work in process (WIP), which is a major source of waste in a development project.

Model-Driven Development (MDD)

MDD uses modeling techniques, for example, use case or unified modeling language (UML) diagrams, to define the business requirements and then aid—and to some degree automate—the generation of code for the intended software application. The models represent an abstraction of the business and user requirements. The automation components of MDD leverage a domain-specific language (DSL) to generate the code. Testing can also be automated against the requirements generated in the models. In effect, MDD is a modern version of RAD.

Short development iterations can be supported leveraging Agile model-driven development (AMDD) tools and techniques, enabling rapid changes in an Agile-based development project. The advantages of MDD and AMDD approaches are rapid development and testing and near defect free code. The concerns about MDD have to do with limitations in the model and automation components to provide enough flexibility to support the needs of the customers and project team.

FEATURE-DRIVEN DEVELOPMENT (FDD)

FDD is considered a lightweight Agile-based development methodology that incorporates OOD, IID, and MDD concepts. Lightweight simply means the methodology is not encumbered with a great deal of bureaucratic overhead, documentation, and structure. The hallmarks of a lightweight methodology are flexibility, speed, and a code-oriented approach as opposed to documentation and a process-oriented approach.

The FDD methodology was initially conceived in 1997 by Jeff De Luca to reduce management overhead, time, and money on a software development project he was involved with in Singapore.[41] A description of the FDD process became available to the public in a book titled *Java Modeling In Color With UML* that De Luca co-authored and published in 1997. There are five steps in the FDD methodology:

- **Develop an overall model**—The scope of work is developed through models of the affected domains, then peer-reviewed and merged.
- **Build a feature list**—Customer-desired capabilities are documented as features in the form of "<action> <result> <object>;" for example, <display> the <value> of an <order>. Each feature is limited in scope to no more than an estimated two weeks of work to create.
- **Plan by feature**—The list of features is planned, scheduled, and assigned to developers; each feature becomes a class within the OO development paradigm.
- **Design by feature**—For each two-week development iteration, a design package is created by the lead programmer that includes the sequence diagrams (e.g., object interactions over time with messaging sequences), plus a description of the implemented classes and methods. A peer-reviewed design inspection occurs before the designs can move into development.
- **Build by feature**—The assigned class owners code the features, ensure that the functionality is implemented correctly via unit testing, and then integrate their code with the mainline code.

Six Sigma

Six sigma employs a statistical approach to process improvement. The term six sigma was coined by a Motorola engineer, Bill Smith, and the term remains a federally registered trademark of Motorola. The objective of six sigma processes is to reduce defects to less than 3.4 per million opportunities. Both lean and six sigma use similar methods and tools. However, where lean has a focus on eliminating waste to improve operational efficiencies, six sigma techniques help eliminate defects and reduce variability in the execution of processes or the quality of products.

Statistical quality control (SQC) techniques are used to indicate whether a process or product is under control in a series of Shewhart/PDSA cycles. Walter Shewhart is often referred to as the father of SQC due to his leading-edge work in statistics and quality control in relation to the control and continuous improvement in processes and products.

Shewhart's work demonstrated that a controlled process requires correction when it deviates three sigma from the mean (e.g., having no more than 0.27% deviation in the upper and lower control limits from the acceptable mean value). In a process that is controlled around a mean following a bell curve distribution, 99.99% of the values will take place within plus or minus three sigma. In other words, there should be only one failure out of every 10,000 executions of the process.

Modern six sigma process improvement techniques are based on Shewhart and Deming's PDSA cycles. There are two forms of this cycle, as shown in Table 12. The DMAIC approach is used to improve an existing business process while the define, measure, analyze, design, and verify (DMADV) approach is used when creating new service, product, or process designs.

Table 12. Six Sigma Techniques

Define, Measure, Analyze, Design, Verify (DMADV)	Define, Measure, Analyze, Improve and Control (DMAIC)
Define	Define
Measure	Measure
Analyze	Analyze
Design	Improve
Verify	Control

Applied to a product's life cycle, DMADV processes tend to be applied across product concept, design, and development activities, while DMAIC techniques find usefulness in the product's operations cycle. See Figure 11 for a conceptual view on how DMADV and DMAIC techniques are applied to major processes within the product life cycle.

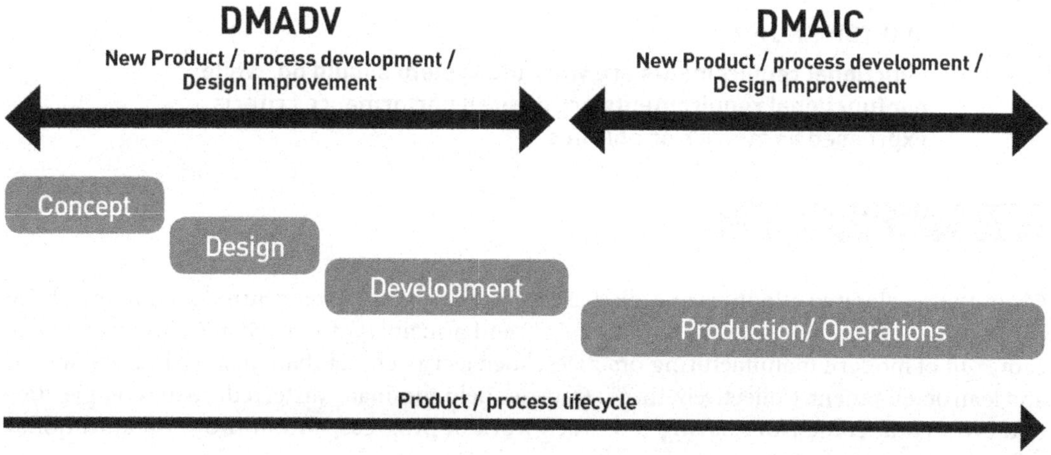

Figure 11. Application of DMADV and DMAIC Improvement Techniques across a Product Life cycle

SIX SIGMA APPLICATIONS IN SOFTWARE DEVELOPMENT

For most of us, it's probably hard to imagine running a software development activity 10,000 times to see if it is out of control. Maybe that's possible in monitoring the CI and CD processes associated with a DevOps environment supporting a large-scale software development project. However, for the most part, the application of six sigma DMAIC techniques is probably not an appropriate measurement for monitoring and controlling front-end software development activities.

However, in IT operations, DMAIC techniques can be applied to ensure that the software or system is stable and performing to expectations. While we don't expect the software code to fail, it still has to support dynamic business processes and work with data that is subject to corruption, and with infrastructure components (e.g., computing hardware, security, and network systems) that are subject to failures. Enterprise software systems that are supporting hundreds or potentially millions of end users, and performing millions of instructions per day, and operating in environments with thousands of virtualized or dedicated computer servers, make DMAIC a useful monitoring and control technique.

On the front end of software development, useful six sigma techniques have evolved in the form of design for six sigma (DFSS). The goal of DFSS is to facilitate discovery of customers and business needs and using that information to derive designs that meet expected performance criteria.

DFSS techniques evolved with an understanding that optimization of products and processes to three sigma standards were nearly impossible without designing the products and processes up front to support those levels of quality. They are improving a process or product after launch is certainly less efficient if not impossible after deployment. Six sigma levels of performance in an operating environment don't happen by accident; it must be built in.

The traditional DMADV model can be applied to software and systems designs. However, there is not a universally recognized process or approach to DFSS. A simpler view of the model might focus on the following elements:

- Identify the functional and nonfunctional requirements
- Design software and systems to implement the desired functional requirements
- Enhance the design to ensure the achievement of nonfunctional requirements
- Verify/validate production operations to confirm conformance with requirements and acceptance criteria and identify areas for improvements

DID YOU KNOW?

Functional requirements are what the system should do, while nonfunctional requirements are desired performance criteria expressed as metrics or abilities.

CHAPTER SUMMARY

Given the accelerated adoption of Agile-based practices, this chapter introduced many of the concepts to development that support the values and principles of agile. Some of these concepts came out of modern manufacturing practices, such as kaizen, Kanban, just in time, six sigma, and lean development. Collectively, these strategies help eliminate waste, reduce the complexities and errors that come from having too much work in progress, and improve organizational efficiencies and throughput.

Other concepts, such as DevOps, had their origins in finding ways to better align software development and support organizations to reduce cycles times-from product concept through delivery, and improve the quality and fitness of delivered products. A DevOps strategy seeks to integrate product life cycle processes and simultaneously improve communications and collaborations between these two traditionally siloed organizations.

With the integration of SDLC processes came the opportunity to gain further efficiencies through automation. Integrated tools supporting test driven development, and continuous integration, testing, delivery and deployment capabilities help enable SDLC automation.

In this chapter, you have learned how all these concepts and capabilities can support the refinement of an agile implementation strategy.

04.
Representative SDLC Methodologies

This chapter offers a cursory understanding of selected SDLC methodologies that integrate one or more of the foundational SDLC concepts introduced in chapter 2 (spiral model, V-shaped model, DSDM, XP, Scrum, DAD, and SAFe).

CHAPTER OBJECTIVES

At the end of this chapter, the reader should be able to discuss the following:

- Describe common SDLC approaches and how they are different from each other.
- What is model driven development?
- List some of the differences between RUP and AUP.
- How does a Cynefin Framework support knowledge management and systems development efforts?
- What makes the Disciplined Agile Toolkit a different approach than other SDLC Frameworks?
- Which modern SDLC methodology encourages the use of development tools in agile based systems development projects? • What are the embraces 5 values, 4 roles, and 12 core practices of Extreme Programming?
- What method was developed to assess government contractors' capabilities to implement a contracted software project based on the maturity of their processes?
- Which set of processes were developed in support of CMM efforts at CMU/SEI?
- Discuss variant approaches to agile processes, such as XP, DSDM, Scrum, SAFe®, and Spiral.
- Explain how the V-Model is similar to and yet different from the traditional linear-sequential SDLC models.

AUTHOR'S INTRODUCTION

At the beginning of Chapter 2, I noted that I do not intend to be an advocate or promote commercialized SDLC methodologies. However, you may be involved in projects that use some of these practices or hear of developers who speak of them. It's important that you see how these various SDLC methodologies build from the fundamental SDLC concepts introduced in Chapter 2.

As you read this chapter, bear in mind that my goal is not to try to make my readers experts on the SDLC methodologies introduced here. Rather, it is to provide a brief introduction and a launch pad to further reading for those who require a deeper understanding of any of the approaches presented.

Diverse SDLC Approaches

There are numerous systems development methodologies and even more technical disciplines and engineering practices that can support a systems development life cycle. The subsections that follow introduce a number of methodologies that support systems development, many of which are still in use today. There are many more systems development methodologies that are not discussed in this section. Again, my intent is not to be an advocate of a particular commercialized approach or try to document every approach you might be exposed to. Rather, the goal of this book is to give you a solid foundation of SDLC methods, processes, and practices upon which you can build.

Software Development and Documentation through Models

Modeling techniques for software development were created to be aids to define sets of best practices for understanding requirements and certain design and development techniques. For example, models were defined to help developers understand information and data relations, hierarchical or functional relationships, OO designs, processes and workflow descriptions, and ontologies–methods to describe semantic meanings and contextual use of words within specific domains, and other techniques.

Early examples of modeling techniques in the software industry, and specifically related to database designs, were the *relational model*, developed by E.F. Codd in 1969, who was then with the IBM Research Laboratory in San Jose, CA,[42] and the *entity-relationship modeling technique* developed for database design by Peter Chen, who was then an Assistant Professor at the MIT Sloan School of Management.[43]

The 1970s and 1980s saw rapid evolution in software and data modeling techniques. One of the earliest efforts to codify and integrate the emerging techniques was by the U.S. Air Force within their Integrated Computer-Aided Manufacturing (ICAM) Program, released as he ICAM Definition, later shortened to IDEF. In 1995, Dr. Richard Mayer, under contract to the Armstrong Laboratory, Logistics Research Division of the U.S. Air Force at Wright-Paterson AFB, formalized and released six IDEF methods.[44] All six of the IDEF methods are in the public domain and available for download as part of the Information Integration for Concurrent Engineering (IICE) Compendium of Methods Report. [45]

I was fortunate to work with in collaboration with Dr. Mayer and his company, Knowledge Based Systems (KBSI), and AT&T Istel, to integrate KBSI's IDEF-based tools with AT&T's simulation products to form an Integrated Computer Aided Software Engineering (ICASE) environment supporting the IDEF methodologies. I was both a sales manager and product manager for this suite of tools before moving over to AT&T Global Information Systems Worldwide Professional Services Divisions as the Director of Methods and Tools for Business Consulting Services.

IDEF primarily remained a modeling technique employed within various federal agencies, such as the Defense Information Systems Agency (DISA) and the Department of Defense (DoD) in general. However, in commerce, the techniques were largely supplanted by the methods and tools supporting the UML and the various UP methodologies and tools.

The terms CASE and ICASE tools were popular for a moment, as so many things tend to be in the IT industry. However, the concepts of having an integrated set of methods and tools for software development live on in the form of modern RAD tools and IDEs.

Both RAD and IDE tools help make software application development activities more efficient. The primary difference between RAD tools and IDEs is their intended users. RAD tools support domain user involvement in application development by minimizing coding. In contrast, IDEs specifically support the code development and testing tasks of computer programmers.

Agile Unified Process (AUP)

AUP, developed by Scott Ambler, is a simplified version of RUP.[46] The approach applies Agile techniques including test-driven development (TDD), Agile model-driven development (AMDD), Agile change management, and database refactoring to improve productivity.

In addition to developing the AUP methodology, Scott has written several books and white papers on OO software development, software process, DAD, AMDD, Agile database techniques, and the Enterprise Unified Process (EUP)™.

Cynefin Framework

Cynefin is not an SDLC framework. Rather, it is a decision-making framework to analyze and solve different types of organizational, policy, and systems problems. A discussion on the Cynefin framework is included in this book for those who have an interest in knowledge management, or its inclusion as a decision support method in other software development strategies.

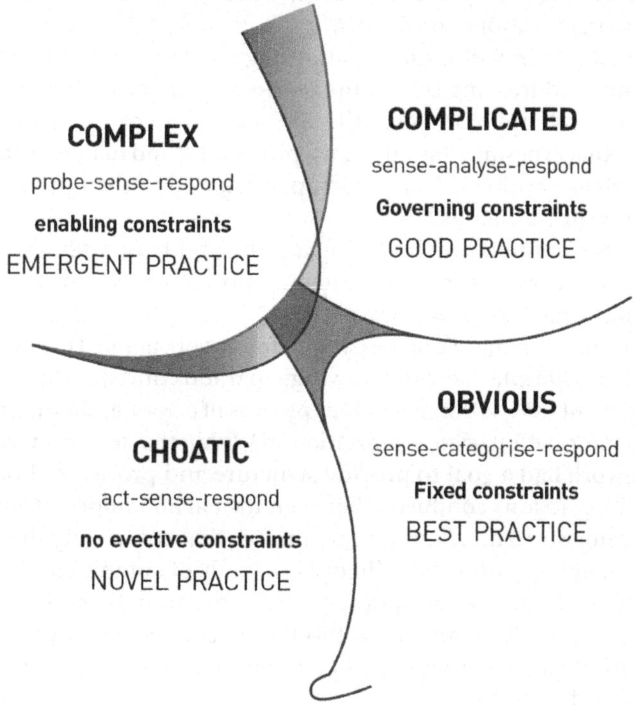

Figure 12. The Cynefin framework - A sense making device

The Cynefin framework provides a quadrant view of four domains with an area of disorder in the middle. Figure 12, Liminal Cynefin, provides a graphical view of these domains: Disorder in the center, Complicated and Obvious to the right, and Complex and Chaotic to the left. The domains on the right represent areas of order, while those on the left represent areas of disorder. The central domain of disorder indicates a state of not knowing which state we are in.

While at first glance it may seem practical to address a problem by identifying and applying known best practices, that may not always be the best approach. If the subject domain is well-known and characterized by a high degree of order and certainty, independent of whether the system is characterized as obvious or complicated, then best practices may very well work. However, if a domain is characterized as having unordered complexity or chaos, then we likely need to explore the root causes before attempting to find a working solution. We may also want to use experimentation and trial and error approaches to seek working solutions before implementation. We may have to come up with an entirely novel solution. The Cynefin framework provides an approach to make such assessments.

The Cynefin framework was created in 1999 by David Snowden, a Welsh management consultant and researcher in the field of knowledge management, who was then working at IBM Global Services. Cynefin is derived from Welsh words meaning "habitat" or "place" but has a broader contextual meaning that implies that our present place is rooted in our past experiences. Put most succinctly, all human interactions are affected by our past experiences, as influenced by geography, religion, culture, education, and other external environmental factors.

In "The New Dynamics of Strategy: Sense-Making in a Complex and Complicated World," written by Cynthia Kurtz, a fellow IBM researcher, and David Snowden, Cynefin is introduced as a sense-making framework "to consider the dynamics of situations, decisions, perspectives, conflicts, and changes in order to come to a consensus for decision-making under uncertainty." [47] The Cynefin framework is described as a sense-making process to utilize the collective experiences of people working within an organization to both understand and solve problems. The framework also helps explain the rationale behind the evolution of complex systems where uncertainty is a major factor.[48/49]

When I asked David Snowden for his input on this section of the book, he was kind enough to review the content and provide some clarifications. One comment he made was to point out that "the liminal domains and dynamics that have been used to understand SDLC and its use are not confined to organizational problems."

Liminal is defined as a transitional or initial stage of a process. When Snowden speaks of Liminal Cynefin, he is addressing the boundary areas of disorder between complicated and complex, and between complex and chaotic. (See shaded areas in Figure 14.) In his writings, Snowden points out that experimentation, pilots, prototypes, and iterative experiments, including methodologies such as Scrum and lean startup, work well in the liminal area bordering the complex and complicated domains.[50]

Snowden originally defined sense-making as "the ability or attempt to make sense of an ambiguous situation."[51] However, in his note to me, Snowden says he now defines sense-making as "How do we make sense of the world to act in it."

Snowden's first writings on the use of the Cynefin framework applied to knowledge management was in a paper titled "Complex acts of knowing—paradox and descriptive self-awareness."[52] Knowledge management (KM) is defined as the "process of creating, sharing, using and managing the knowledge and information of an organization."[53] Within the field of knowledge management, the Cynefin framework had a goal to provide structure and process to both differentiate and communicate the interactions conducted between formal and informal communities.

In his direct comments to me, Snowden points out that complexity theory is a naturalizing approach to sense-making. Complexity theory is a study of complex and chaotic systems and how order, pattern, and structure can spontaneously arise from them. In other words, complex systems are adaptive and self-organizing without the need for some type of master controller. It is the interactions of participating agents acting in self-interest that leads the system out of chaos and into ordered patterns.

Disciplined Agile (DA)

Throughout my career, I have supported various customers and employers who have enforced the use of waterfall, iterative waterfall, XP, hybrid, Scrum, and DevOps-based SDLC methodologies on their IT projects. As a contracts-based IT program and project manager, I don't have much influence on which SDLC approaches are used, as the client's IT governance policies dictate those decisions. My job is to make sure the corporate standards governing the approved methods and tools are appropriately understood and utilized by the project's development teams.

DID YOU KNOW?

In his review of this section, Scott Ambler noted that the DA community of practitioners are moving away from the term "framework" in favor of the phrase "toolkit."

When I started writing this book three years ago, I was coming off a large health IT-related program (encompassing 16 parallel projects and ~100 participating program team members) that leveraged a hybrid waterfall-Agile based SDLC model. That program heavily influenced my thinking, as I liked the idea of tailoring our project to employ the practices that best fit our needs.

Though I was aware that an approach called DAD existed, I didn't have any practical experience with it, as my customers and employers had not implemented this approach in their organizations. However, as I reached out to Scott Ambler to get his thoughts on a different section within this book, he noted that his current work is focused on enhancements to disciplined Agile, and particularly the DAD portion of it. As a result, I spent some time coming up to speed with this SDLC approach. I'm glad I did, as the DAD approach fits very well with my view on what an SDLC toolkit or framework should include.

In collaboration with Mark Lines, Scott Ambler developed the disciplined Agile delivery (DAD) process, now included as part of the DA Toolkit, during his time as chief methodologist for IT at IBM Rational (2006 to 2012). DAD provides a comprehensive framework of Agile-based SDLC processes to overcome the limited scope of other Agile methodologies. For those who want to learn more, Scott and Mark have written four books on DAD.[54/55/56/57]

The Disciplined Agile Consortium (DAC) is the governing body for certification in the DA Toolkit. DAC promotes DA as a foundation for business agility. [58] DAD in its current form is a hybrid framework that provides a full life cycle set of processes that expand beyond the traditional boundaries of Scrum and other development-focused Agile methodologies. The objective of DAD is to provide guidance on best development approaches under different situations and not prescriptive directions that lack the flexibility to address unique customer, product, project, or program needs.

DID YOU KNOW?

Scott Ambler has led the development of several software processes, including DA (with Mark Lines), AM, AD, EUP, and AUP methodologies.

DAD is also referred to as a lightweight SDLC model. However, it is not lightweight because it lacks a comprehensive set of SDLC methods and tools. The DA Toolkit is constructed to incorporate a plethora of SDLC practices, including Agile data (AD), Agile Modeling (AM), Cynefin, DevOps, lean software development, Kanban, SAFe, Scrum, XP, UP, and traditional software practices. The framework supports both lean and Agile ways of working and was built using empirical evidence of best practices obtained from interviewing hundreds of organizations.

While the DA Toolkit supports a full set of product life cycle processes, from conception through end-of-life, it is laid out differently than a traditional SDLC Framework of phase-based processes. The focus of the DA Toolkit is to implement and integrate lean-Agile practices throughout the product life cycle and across the enterprise. Therefore, the scope of the DA Toolkit now expands out from the original DAD concepts to implement disciplined DevOps, DAIT, and disciplined Agile enterprise (DAE) capabilities.

- **DAD—implementation of core lean/Agile processes across a solution life cycle**
- **Disciplined DevOps—has a focus on streamlining IT solution development and IT operations activities**
- **DAIT—expands lean/Agile strategies and processes across all IT functions**
- **DAE—implements lean/Agile strategies and processes throughout the enterprise to drive innovation and respond to changes in the marketplace**

Through his consulting firm, Ambysoft, Inc., Scott Ambler advanced the various disciplines of "Disciplined Agile Delivery". The methods originally supported include AM, AD, and the EUP.

- AM is a practice-based methodology for effective modeling and documentation of software-based systems.[59]
- AD is a method defining a collection of strategies, techniques, and philosophies, to help teams work effectively on the data aspects of software systems.[60]
- EUP provides an extension to solution delivery methodologies such as DAD, instantiations of the UP, Scrum, and other Agile software methods.[61]

In their new book, "Choose Your WoW!" Ambler and Lines show how an organization can apply the DAD portion of the DA Toolkit via guided continuous improvement decisions. (e.g., along the lines of Kaizen but incorporating team process expertise to guide their decisions).[62] Their new book also demonstrates how to select from one of the six DA life cycles and then choose appropriate practices via process goal diagrams. Organizational practices evolve as the organization continues to experiment and learn.

Dynamic Systems Development Model (DSDM)

DSDM is an Agile-based development framework initially supporting software development projects but subsequently evolved to be more generic in use. DSDM was created in the United Kingdom in the 1990s by a consortium of vendors and experts in information technology, originally known as the DSDM Consortium. This organization was later renamed the Agile Business Consortium.[63] DSDM remains vendor-neutral and covers the entire life cycle development process.

The initial goal of the consortium was to address issues that project managers faced using RAD and IDE tools and techniques in Agile-based projects. In other words, the objective was to leverage the RAD concept on IID projects in support of evolving or changing requirements.

The current stated focus of DSDM is to provide an approach to development that delivers products on-time, in-budget delivery of projects, with proven scalability to address projects of all sizes. DSDM is built on the philosophy that all projects must be in alignment with the organization's strategic goals and provide tangible benefits to the business.

DSDM advocates the use of proven practices in the form of:

- Facilitated workshops
- Modeling and iterative development practices
- MoSCoW (must have, should have, could have, and won't have) prioritization
- Timeboxing

eXtreme Programming (XP)

This was the first Agile-based SDLC approach that I was involved with. The success of the software development projects, particularly in terms of customer and end user acceptance, was so positive that I truly did not think I'd ever manage another traditional waterfall-based project in my career. Unfortunately, I had future customers who had different ideas about that, usually for the wrong reasons. Since I have personal experience with this methodology and success with it, I'll spend a bit more time in this review.

Kent Beck created XP during his work on the Chrysler Comprehensive Compensation System (C3) payroll project.[64] Kent Beck is one of the signatories on the Agile methodology, and he remains one of the top influencers within the IT industry.[65]

The XP methodology embraces 5 values, 4 roles, and 12 core practices. The 5 values are:

- Communication—with a focus on knowledge transfer
- Simplicity—do only necessary things
- Feedback—to identify areas for improvement and optimization
- Courage—Prefer actions that aren't harmful to the team
- Respect—Provide feedback that honors your relationship to other team members

The 4 roles documented within the XP methodology are:

- Tracker/manager—Overall responsibility for defining rules of the planning game, release and iteration planning, and tracking core metrics
- Customer—Somewhat similar to the product owner role in Scrum, but more of an influence on the development team than a director. Also involved in documenting requirements via user stories and writing related functional test scripts.
- Developers/programmers/testers—Considered the most important role in XP, these individuals have skills and responsibilities for estimating development tasks, developing code, writing and executing unit tests, refactoring code, and integrating code
- Coach—Provides instruction and guides the team on the proper use of the XP methodology

The 12 core practices are:

- User stories for requirements gathering
- Small iterative code releases
- Common naming conventions and descriptions
- OO development techniques
- Paired programming (e.g., two programmers sitting side-by-side in collaboration)
- Automated unit testing
- Refactoring
- Continuous code integration
- No overtime
- Full-time customer representation
- A common coding standard

XP has proven to be a viable Agile-based approach for many organizations. This isn't to say that the XP approach did not have any downsides. For example, scope creep can still set in if boundaries are not placed on what's in and what's out of a funded and approved project cycle. There are concerns that developers may place too much emphasis on coding before the architecture and designs are properly fleshed out. The lack of documentation, particularly on acceptance criteria related to requirements, can lead to bugs and defects in the code. Finally, the methodology doesn't scale well on large enterprise projects, particularly those involving many developers working across disparate sites

On the other hand, an XP-based development team does not need to explicitly follow all 12 core practices. The facilities or use of remote teams can affect decisions on the ability to implement paired-programming—and in my experience, I've found that limitation can be overcome with online chat, instant messaging, and screen sharing tools.

Also, it can be quite a challenge to get a customer to commit a full-time and qualified domain expert to sit on-site with the development team all day. Our teams were able to overcome those challenges with iterative product reviews. However, you may have a customer who is unwilling to encourage/mandate that staff members and other domain experts participate in periodic product demos and requirements-gathering sessions.

Though other Agile-based methodologies such as Scrum, Kanban Scrumban, and DevOps, have become more popularized, XP remains a viable and preferred approach in some organizations. In my opinion, it provides a relevant advantage on smaller projects where speed and flexibility are valued, and where the lightweight methodology and its relative simplicity do not introduce overwhelming complexity or logistical burdens.

Open Unified Process (OpenUP)

OpenUp supports the unified process as an open source component of the Eclipse Process Framework (EPF), maintained under the Eclipse Foundation. OpenUP is promoted as an iterative software development process that is minimal, complete, and extensible. The OpenUp approach supports team collaboration and focuses on stakeholder benefits while minimizing the opportunity for fruitless deliverables and excessive formality.

OpenUP is driven by the following core principles:

- **Collaborate to align interests and share an understanding**
- **Balance competing priorities to maximize stakeholder value**
- **Focus on the architecture early to minimize risks and organize development**
- **Evolve to obtain feedback and improve continuously**
- **Eclipse is an open-source IDE used in computer programming. The Eclipse Foundation provides their global community of members, both individuals and organizations, with a mature, scalable, and commercially friendly environment for open source software collaboration and innovation.**

Though created by IBM in November 2001, the Eclipse Project was supported by a consortium of software vendors from the beginning. The Eclipse open source project was reformed as the Eclipse Foundation in January 2004 as an independent, not-for-profit corporation to provide ongoing stewardship of the Eclipse community. The Eclipse Foundation supports both commercial and open source projects within the member communities that adopt the Eclipse technology.

Outside-In Software Development

The outside-in software development methodology is considered an Agile-based development technique. The name "outside-in" is derived from a philosophy that the requirements of stakeholders "outside" the development team drive all development activities "inside" the team. The outside-in software development methodology is not considered a standalone SDLC. Rather, it is used to augment other Agile/lean approaches to ensure that the development team stays focused on the needs and motivations of the stakeholders who use the final solutions. The outside-in approach identifies four key stakeholder groups:

- **Principals—Those who are paying for the software, since they affect funding, resources and schedule, and can kill a failing project, are the most important stakeholders to please.**
- **End users—The individuals who use the product. They are also the folks who are most likely to understand what capabilities must be delivered with a product to ensure its usefulness.**
- **Partners—This category of people includes IT operations (support and maintenance groups), other development teams and systems integrators, and external partners affected by the software or system under construction.**
- **Insiders—In my opinion, these are the people who make up the original definition of stakeholder (anyone within or external to the organization who has an opinion that matters). Not only can insiders drive development priorities, if they perceive that the product, team, or use of corporate resources is a threat to their interests, they may work against the development team to undermine the funding and support provided by the Principles.**

An early book on this subject was written in 2007 by Carl Kessler and John Sweitzer, titled: *Outside in Software Development: A Practical Approach to Building Successful Stakeholder-based* Products.[66] Carl Kessler is vice president of worldwide development within IBM's Software Group. John Sweitzer is an IBM Distinguished Engineer and a member of IBM's Academy of Technology. Their book describes five steps to outside-in thinking, plus their thoughts on how to initiate an outside-in approach to development.

The primary premise of the outside-in philosophy is that software development activities and enhancements taken to support corporate process, policy, people, or systems cannot be the driver. Instead, the organization needs to turn their thinking around and understand that maximizing customer experience, customer benefits, and customer interactions must drive development priorities. A focus on improving external customer and stakeholder benefits will achieve organizational benefits that transcend those that can be obtained when strictly following inward directed objectives. [67]

CMM

A CMM is not an SDLC process; rather, it is a formalized approach to assess a software development company's development and acquisition processes. The U.S. Department of Defense contracted the Carnegie Mellon University (CMU) in 1987 to conduct related research and develop a CMM. The initial purpose of CMM was to describe a method to assess government contractors' capabilities to implement a contracted software project based on the maturity of their processes.

Watts Humphrey founded the Software Process Program at CMU's Software Engineering Institute (SEI) and is credited with overseeing the development of the Capability Maturity Model. Some have referred to Humphrey as the "father of software quality." Humphrey's efforts at CMU also led to the development of the personal software process (PSP) and the team software process (TSP).

The first version of CMM was developed and released in 1991 as the software development capability maturity model.[68] The term "maturity" represents the degree to which an organization has formalized and optimized their business processes. CMM implements a set of processes where maturity is achieved by an organization through a series of four stages:

1. Ad hoc practices
2. Formal definition and documentation of process steps
3. Controlled process management to desired result metrics
4. Processes optimization

CMM was officially sunset by the SEI on December 31, 2005, supplanted in 2006 by the SEI's improved CMMI.

CMMI

CMMI version 2.0 is a process-level improvement training and appraisal program that originally came out of the same CMM effort contracted by the DOD to CMU SEI. The original CMMI product team included process improvement experts from the government, industry, and the SEI, with the goal to improve on the existing software CMM.

In 2012, the CMMI research, training, and information responsibilities were transitioned to The CMMI Institute, then a wholly owned subsidiary of CMU. However, in March 2016, the Information Systems Audit and Control Association (ISACA) purchased the CMMI Institute.

The CMMI Institute retains ownership rights over all intellectual property and operates as an independent entity to administer the CMMI-based process improvement training and appraisal programs. CMMI version 2.0 is the current release of CMMI.

CMMI version 2.0 provides improvements in four major areas: a new focus on improving organizational performance regarding ROI; support for organizations using Agile-based development practices; improvements in CMMI usability supporting wider accessibility and comprehension; and improvements to the SCAMPI and Benchmark appraisal systems.

CMMI defines the following six maturity levels for processes within the organization:

- **Incomplete**—Ad hoc and unknown processes are used within the organization
- **Initial**—Unpredictable and reactive processes often cause delays and budget overruns
- **Managed**—Project level management that is planned, executed, measured, and controlled
- **Defined**—Guidance enables proactive rather than reactive use of standards across projects, programs, and portfolios
- **Quantitatively Managed**—Organization actively collects data to compare against quantitative performance objectives that are predictable and in alignment with stakeholder needs
- **Optimizing**—Processes are stable and flexible, and the organization is focused on obtaining continuous process improvements

The CMMI Institute continues to develop maturity programs beyond the scope of software and systems engineering practices to facilitate quality maturation of company products and services regardless of size or type of industry. While the CMMI Institute's focus remains on benchmarking and improving business capabilities, their scope has expanded to include product development, service excellence, workforce management, data management, supplier management, and cybersecurity practices.

In addition, the CMMI Institute has developed new maturity programs, such as the Data Maturity Management (DMM) model introduced in 2014. In 2017, the CMMI institute partnered with the federal government on two healthcare initiatives to improve patient safety: the Medical Device Discovery Appraisal Program (MDDAP), and the Patient Demographic Data Quality Framework (PDDQ Framework). In 2018, the CMMI Institute introduced CMMI version 2.0.

Having been involved in government contracting for a significant portion of my career, I've personally supported several CMMI appraisal preparation activities by the firms I worked for. The trick is sustaining the activity in the long run. It's not that the effort is not worthwhile; the problem is that people within the organizations have regular jobs to do, and it can take a significant amount of time and effort to support a CMMI appraisal effort. If the CMMI programs and appraisals are viewed to have a negative impact on the focus of the business, management won't support the CMMI initiatives. Put another way, executive management must commit enough time and resources to achieve the benefits of a maturity program and make the work a priority.

PSP and TSP

The product PSP and TSP were both developed by Watts Humphrey as part of the CMM efforts at CMU/SEI. Watts believed that large-scale and highly complex systems could be built with high quality and reliability by highly skilled technical teams using rigorous development processes.

PSP provides a structured software development process that helps software engineers understand and better improve their performance. In contrast, TSP provides a guide for engineers and managers to form self-directed teams and produce high-quality software on schedule and budget. The self-directed teams manage themselves, plan and track their work, manage the quality of their work, and work proactively to meet team goals.

The SEI claims that "Since the early 1990s, thousands of engineers at organizations worldwide have used the PSP and the TSP to produce nearly error-free software programs."

A lot of research and documentation is available on the PSP/TSP methods. For those who want to read further on these subjects, the CMU/SEI maintains a "Bibliography for the PSP and TSP" on their website. [69] The SEI also maintains an extensive library of content on the PSP-TSP section on the SEI's Software Engineering Information Repository (SEIR). [70] There are also books written on the subject, including two written by Watt's Humphrey.[71/72]

RUP

RUP is an iterative software development process framework originally created by the Rational Software Corporation. Rational was acquired by IBM and has been a division within IBM since February 2003. The Rational Environment is first and foremost an IDE. Secondarily, the Rational tools implement IBM's latest views on software development processes.

In the early to mid-1990s, Rational acquired companies that brought three of the leading thought leaders in OO programming together into one company, Grady Booch, James Rumbaugh, and Ivar Jacobson. Each of these individuals had defined OO-based development approaches that were separate and unique: the Booch method, the object-modeling technique (OMT), and object-oriented software engineering (OOSE).

Under the Rational umbrella, they refined and integrated their approaches into a single language called UML. In 1997, UML was adopted as a standard by the Object Management Group (OMG), and OMG has maintained and updated the UML process since.

Based on UML, RUP was a specific implementation of the Unified Process on the IBM Rational platform, and specifically on the IBM Rational method composer (RMC) product. The RUP development approach supported iterative development, component-based architectures, modeling, continuous developer-driven testing, requirements management, and automated testing. IBM later released RUP as the open source method called OpenUP through the Eclipse process framework (EPF).

SAFe®

SAFe is a development framework based on Lean, Agile, Systems Thinking, and DevOps practices, but with the specific objective to scale these capabilities on an enterprise scale. The SAFe framework is developed, maintained and enhanced by Scaled Agile, Inc. Their latest version is SAFe 5.0, which provides a knowledge base of proven integrated principles, practices, and competencies to achieve business agility.

The SAFe framework offers four configurations:

- Essential SAFe
- Large Solution SAFe
- Portfolio SAFe
- Full SAFe.

Essential Safe begins to scale Lean-Agile practices at the small group size of 50 to 150 people, is the most basic configuration of the SAFe framework, and implements the smallest subset of elements necessary to successfully implement SAFe. *Large Solution* SAFe provides additional guidance to organizations that build large and complex solutions. *Portfolio SAFe* implements practices to align portfolio strategies, investment funding and governance to build the right set of products that maximize enterprise value. Finally, *Full Safe* is the most comprehensive implementation of SAFe that includes all seven core competencies defined by the company as necessary to achieve Business Agility:

- Team and Technical Agility
- Agile Product Delivery
- Enterprise Solution Delivery
- Lean Portfolio Management
- Organizational Agility
- Continuous Learning Culture
- Lean-Agile Leadership

There are other approaches to scaling Lean and Agile concepts across an organization. But one of the unique aspects of Scaled Agile's approach is their detailed Implementation Roadmap. Instead of leaving it up to companies to figure out how to implement Lean-Agile practices at scale, the SAFe implementation roadmap offers a comprehensive set of events, roles, artifacts and training courses to guide the effort.

SAFe practices were originally refined and published in Dean Leffingwell's book: "Scaling Software Agility: Best Practices for Large Enterprises.[73] Dean Leffingwell remains involved with developing SAFe® as the Chief Methodologist at Scaled Agile, Inc. The organization cites several case studies that demonstrate improvements in productivity, quality, time-to-market, and employee engagement and job satisfaction. Scaled Agile company continues to offer new upgrades to the SAFe framework. [74]

Scrum Agile Methodology

Scrum is one of the most recognized and acclaimed approaches to implementing Agile-based development practices. I use Scrum in my book titled *Frameworks for Agility* as the reference model for implementing Agile practices within a hybrid SDLC framework, so I won't spend much time talking about it here.

The term Scrum predates its use as a software development strategy. The term was first coined by Hirotaka Takeuchi and Ikujiro Nonaka as an analogy for product development in their 1986 *Harvard Business Review* article, "The New New Product Development Game."[75] The approach was based on the case results of studies from manufacturing firms in the automotive, photocopier, and printer industries.

They viewed successful development practices as a process performed by one cross-functional team across multiple overlapping phases. The team as an organized unit is responsible for going the distance in delivery of the final product, but they pass the workload around among themselves based on skills and needs through the life cycle of developing and supporting the product. In rugby, the term Scrum implies a restart of play. In software development, the restart occurs daily as each Scrum team conducts a short meeting to review progress, synchronize activities, and create a plan for the next 24 hours. In a larger sense of the word, the Scrum methodology emphasizes daily communications, team flexibility, and product development through a series of relatively short iterations of work.

The Scrum methodology was separately developed by Ken Schwaber and Jeff Sutherland in the early 1990s, and then later described and published in a paper describing the Scrum framework at a Business Object Design and Implementation Workshop at an Object-Oriented Programming, Systems, Languages & Applications '95 (OOPSLA) conference in Austin, TX, in 1995.[76]

Today, Ken Schwaber and Jeff Sutherland have formalized the Scrum methodology in the form of a Scrum Guide. The Scrum Guide can be downloaded from their Scrum Guides website.[77]

Spiral SDLC Model

The Spiral type of development model is another early example of employing IID and prototyping techniques within systems development. RAD as a term was not coined until 1991 by James Martin. However, the Spiral methodology depended heavily on the use of 4th generation software development tools and IDEs. 4th generation programming languages evolved an evolution away from procedural programming languages to computer language programs that incorporated higher abstractions in the form of English-like words and phrases.

The Spiral model was initially described by Barry Boehm, then employed by TRW Defense Systems Group, in a paper he published in 1986 titled, "A Spiral Model of Software Development and Enhancement."[78] Barry Boehm published a second paper with an identical title in 1988. "[79] Boehm noted that his model took a "risk-driven approach" as opposed to the document driven or code-driven processes" found in other SDLC methodologies. Boehm believed a major problem with traditional waterfall-based development practices was the stress placed on documenting requirements early on, which then served as completion criteria for the remainder of the development life cycle. This approach, he argued, did not produce good results in terms of identifying and understanding user interface requirements and decision support needs.

As I've noted in previous sections, it's hard for customers and end users to understand what they need, and what is possible, until they have some sort of working prototype in their hands. Barry Boehm's spiral model was an early attempt to address this human factor in software development.

Boehm proposed an approach he named the "evolutionary development model" where interface and decision support criteria are exposed across several incremental developments and delivery cycles. Figure 13 - The evolutionary development model (a.k.a. Spiral Model) provides a graphical depiction of the Spiral model. Boehm's spiral model views development as a continuous and expanding process. The activity begins with an initial hypothesis and risk assessment activity and then spirals outward through any number of prototyping cycles necessary to develop a working software solution.

The spiral model starts with an inner loop (Round 0) that starts with an initial hypothesis about an operational mission (or set of missions) that might be improved with the use of a software product, plus risk analysis to determine the merits of the proposed development effort.

In other words, this round of activity has a focus on evaluating the use of a software solution to address some business or user need.

If the initial hypothesis appears sound, the next phase of the spiral (Round 1 Prototype) begins with a prototyping effort, proceeds with a CONOPS investigation, and produces a requirements plan. Another risk analysis is performed before starting on the next outer loop of processes.

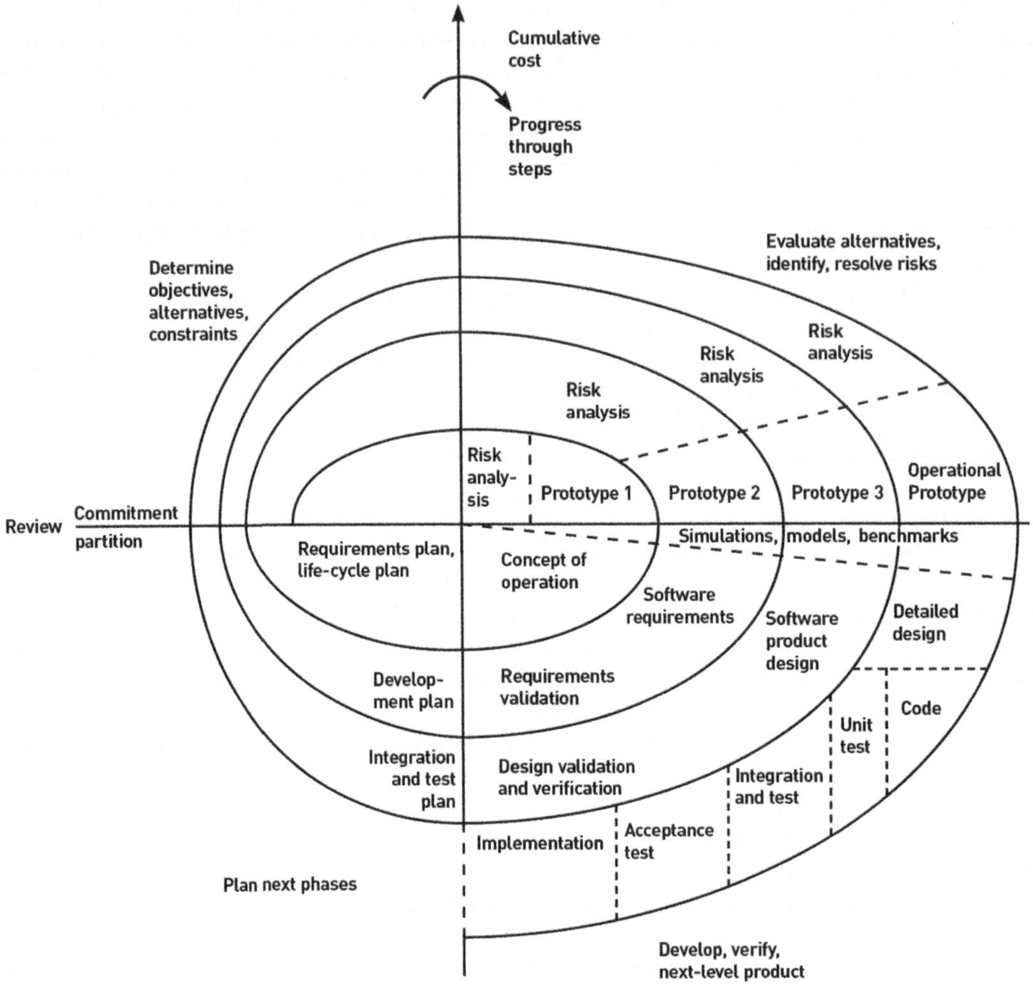

Figure 13. The evolutionary development model (a.k.a. Spiral Model)

This next phase of activities (Round 2 Prototype) includes a prototyping phase that incorporates simulation and other modeling techniques and benchmarks, plus the development and validation of software requirements. The output of this phase is a development plan.

The fourth phase (Round 3 Prototype) again starts with risk analysis, prototyping, and modeling with the goal to form and validate the software product design. The output of the third look is an integration and test plan.

Assuming the project gets this far, the fifth phase (Operational Prototype) again starts with a risk assessment, the development of the operational prototype, and a detailed product design, plus coding, unit testing, integration, and integration testing, UAT, and implementation.

When laid out visually, the continuous process of the evolutionary development model forms a spiral shape. The spiral process can continue for as many iterations (loops in the spiral) as necessary to support the customer's ongoing needs.

V-Shaped SDLC Model

The V-shaped SDLC model (a.k.a. verification and validation model) is a refinement of the traditional waterfall model. The goal of the V-shaped model is to validate each step of the traditional waterfall-based approach through planning and testing before moving on to the next phase in the development life cycle. Truly, these activities should be going on in the traditional waterfall model, regardless. However, the V-shaped model formalizes the sequence.

A German firm, IABG Information Technology, claims they originally developed the V-shaped model for the Federal Ministry of Defense, in cooperation with the Federal Office for Defense

Technology and Procurement. They also note sustainment of the methodology was taken over by the Federal Ministry of the Interior for the civilian public authorities' domain in summer 1992.[80]

The V-Shaped model is a bit more interesting than it might at first appear. First, note that the model follows the traditional waterfall approach to coding, and then stops. Where testing is depicted as a process succeeding the coding process in the traditional waterfall model, test planning is shown as a parallel set of activities in the V-shaped model, with testing following coding tasks but mapped back to the preceding test planning activities. This mapping ensures verification and validation of the code, integrated components, system, and business and user requirements, in that order. In that sense, the V-shaped model is a more accurate representation of how this planning and testing work should occur in a software development project.

Second, while I have viewed numerous graphic versions of the V-shaped model, I prefer the version shown in Figure 14. The left-hand side of the model shows the typical waterfall processes from requirements analysis and related documentation down through coding. The center of the model shows the related test plans necessary to verify successful completion of each step on the left, while the right-hand side of the V-shaped model shows the type of testing that validates successful completion of each stage mapped on the left.

Since requirements must come from customers and end users, and those same folks must validate the fulfillment of the requirements, it makes sense to link user acceptance testing through a user acceptance test plan. Also necessary with the delivery of the user acceptance test plan are test cases, test scripts and test scenarios that map back to the business and user requirements. Test cases indicate the environment that is necessary to test the software. Test scripts are discrete steps or instructions the tester must take to complete a test. Test scenarios are more detailed test scripts that allow the user to test end-to-end functionality of the software in support of a business process or other extended set of linked activities.

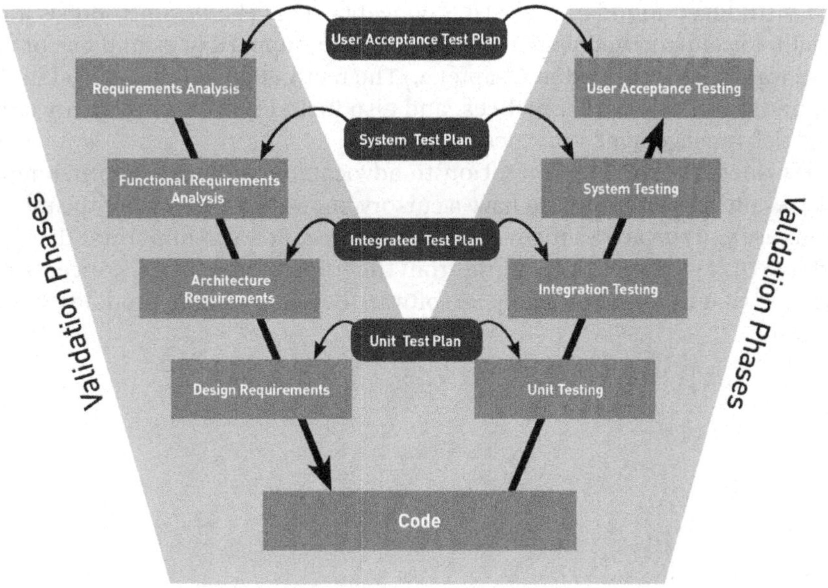

Figure 14. V-Shaped Model (a.k.a. Verification & Validation Model)

A functional specification document (FSD) is produced by the system architect or by members of the development team to describe the desired behavior of the proposed system. While customers and end users typically express their requirements as a set of desired capabilities, the architect or development team must translate those requirements into a language that describes how the system will provide those capabilities (e.g., features and functions, workflows, data flows, and other system level requirements). The system test plan captures these requirements and lays out the tasks the development team must take during system testing to ensure that the product is correctly implementing the capabilities.

The software/system architecture document (SAD), developed by the development team's software or system architect, details the software components necessary to provide the capabilities listed in the FSD. Moreover, the SAD also describes the necessary component interfaces and describes the communications or data exchanges between the components. A SAD typically evaluates system requirements from a number of different views, such as use case, logical, process, deployment, implementation, and data. This subject is revisited in my book titled *Frameworks for Agility*. For now, it's important to note that the SAD drives development of the integration test plan with a description of the components that must be integrated and tested in terms of their interfaces.

The design specification document (DSD) provides a detailed description of how the software components must be constructed. Where the SAD document looks at the software across the breadth of its design needs, the DSD provides a deep dive into how the software or system will work. The DSD is a written guide for the developers who will create and test the software code. The unit test plan identifies the pieces of code that must be developed, the data inputs to test the code, and the expected outputs to validate the code functions as intended. Unit testing is the software validation procedure at this level of the model.

You may wonder why the arrow on the left-hand side of the V-shaped model points down, while the arrow on the right-hand side points up. The answer is straightforward. The work to develop code and build the related test plans proceeds from the top-down on the left and center, while the work to test the software proceeds from the bottom-up on the right-hand side of the model. So, in the end you can think of the V-shaped model as upending the waterfall model at the junction of coding and testing, with testing shown in much greater detail.

CHAPTER SUMMARY

Chapters 2 introduced foundational SDLC concepts. This chapter introduces a number of SDLC and other useful methodologies that incorporate, support, or extend one or more of the foundational practices identified in Chapter 2. The methodologies introduced in the chapter span both traditional and agile practices, and also includes process maturity and decision making and risk management strategies.

This book does not have the intention to advocate or promote commercialized SDLC methodologies. But it's important to have a cursory knowledge on some of the most prevalent practices, in case you run across them during your career. It's also important that you see how these various SDLC methodologies build from the fundamental SDLC concepts introduced in Chapter 2, as that trend will not stop as software development methodologies continue to evolve over time.

05.
Making a Case for the Hybrid SDLC Model

This chapter offers a deeper understanding of how traditional phase or stage-based development life cycle approaches differ from Agile-based development practices, and why a mature SDLC Framework requires elements from both disciplines.

CHAPTER OBJECTIVES

At the end of this chapter, the student should be able to discuss the following:

- Explain why some development teams and customers still prefer the use of traditional plan driven and linear-sequential SDLC Processes.
- Why is some documentation necessary on every software or systems development project?
- Describe the purpose and value of configuration management practices.
- Explain why Agile is a set of values and principles and not truly a specific process.
- Discuss at least three development approaches that support agile values and principles.
- List at least two reasons that support implementation of a Hybrid SDLC Framework

AUTHOR'S INTRODUCTION

I've discussed Agile practices and methodologies extensively in the first four chapters of this book, and how such practices help address some of the issues associated with following linear software life cycle development practices. I mentioned in chapter 2 that the use of waterfall-based practices is not dead. On the other hand, recent data indicates the use of Agile appears to have taken the lead. For example, in the CollabNet/ VersionOne 12th Annual State of Agile™ Report, 52% of the respondents noted that half of their development teams use Agile practices.[81]

Nevertheless, a central theme of this book is that we shouldn't throw out the baby with the bathwater, so to speak, with regard to valuing traditional practices, particularly in the area of engineering. In that context, an SDLC framework helps organize, maintain, and integrate organizational SDLC processes, which can incorporate any number of engineering, development, and operations support methodologies.

The issue I've had with some of the early Agile approaches is that there was a tendency to trivialize the importance of good engineering practice (GEP). I don't think it was intended. Agile practices were largely promoted by developers who recognized traditional software development practices were inefficient, bloated, and not sufficiently helpful in aligning development activities with customer and end user needs.

In my opinion, these developers already understood the importance of GEP, and competence in engineering was a given. Also, it's clear that the authors of the Agile Manifesto had specific preferences in their technical approaches to implementing Agile-based practices. At an organizational level, engineering knowledge is often maintained in the heads of the experts and shared informally among the team in their work settings.

In short, the proponents of Agile were already technology experts in their fields. Their concerns were primarily focused on implementing process-oriented SDLC improvements, not technology-based improvements. Moreover, the emphasis of Agile was on making development less burdensome and more efficient, and not on which technologies, standards, or architectures work best for the organization.

However, technology, enterprise architecture, and related engineering practices do matter and must be in alignment with the organization's IT governance policies and technical standards. Development teams must support the technologies and standards approved in the organization's IT governance policies, enterprise architecture, and technical reference model (TRM). Development teams cannot just use whatever methods and tools they prefer.

There's a simple reason for having and enforcing corporate standards. Every IT system and product become legacy eventually. Development teams that go off the ranch and use whatever technologies they prefer create sustainability issues and higher costs in the future that must support numerous and disparate legacy offerings for the organization.

Yes, organizations adopt new technologies and standards over time, but there are open standards that can ease the transition. Moreover, investments in technology upgrades are business decisions that must be carefully considered from the perspective of risk-reward, the ROI, and supportability and sustainability issues.

Stuff accumulates over time, and the information on how to build, extend, secure, integrate, support, and maintain those systems accumulates. At some level, all this information must be maintained, structured, and made accessible in a logical manner.

I'm not trying to say that corporate IT policies, standards, enterprise architecture, and TRF must be included as components of a corporate SDLC. On the other hand, there is no reason why they can't be. Regardless, the corporate standards drive technology decisions, which in turn drive related methods, tools, and GEP. Corporate knowledge in those areas must be maintained within a corporate SDLC.

My personal view is that the traditional phase or stage based SDLC models provided a serviceable framework for maintaining most SDLC information related to engineering practices. However, that may not always be the best method for categorizing SDLC-related information. For example, some IDEs blur the lines between requirements, design, development, and testing activities. In those cases, a different classification strategy built around the processes of the IDE might make sense. The IT organization also must ensure that the corporate SDLC structure can facilitate the management and integration of GEP with the organization's Agile methodologies.

Some modern SDLC frameworks, such as Disciplined Agile, recognize the value of having an initial inception phase that precedes construction. That doesn't make DAD a waterfall practice. The developers are simply acknowledging that some amount of investigation and analysis is prudent before jumping into prototyping and iterative development cycles.

Depending on the technologies, standards, and methodologies employed, the IT organization needs to figure out a taxonomy that makes sense for organizing SDLC processes approved for use within the enterprise. By now, based on this lengthy preamble, it should be clear that a homogeneous approach to development isn't practical. Organizational SDLC models need hybridization to include engineering practices that support enterprise architecture and technology standards, useful across different types of development projects, plus linking and supporting operational processes.

Hybrid SDLC models are sometimes thought of as an SDLC approach that combines waterfall and Agile-based practices. They certainly can, but I think that misses the point. While I don't see the use of traditional sequential development processes as becoming mainstream again, the fundamental engineering practices and processes don't go away. Therefore, a modern definition for a hybrid SDLC model might suggest a framework that implements a taxonomy or other structure for managing information related to engineering, Agile development, and operations across the life cycle of a product.

Moreover, a modern corporate SDLC might include any number of systems development approaches, technologies, and supporting tools. That is, in fact, the approach of the DA Toolkit. The DA Consortium is moving away from their original view of Disciplined agile as a framework to now referring to it as a DA toolkit. The idea is their SDLC model provides access to a plethora of tools, and the users are free to select those practices that are most appropriate to the needs of their project team.

I discuss the implementation and use of a hybrid SDLC model in my titled Frameworks for Agility. In that book, I spend more time discussing how to integrate disparate processes within an SDLC framework. However, in this chapter, I want to discuss the value of implementing both traditional waterfall and modern Agile processes within a corporate SDLC.

TRADITIONAL SDLC PRACTICES

While the strict implementation of linear-sequential life cycle development processes is a flawed model, some traditional waterfall-based practices are not likely going away anytime soon. For example, the traditional phase-based structure made it easier to break out and document best practices, methods, and tools employed by the organization across the development life cycle.

Also, though the prototyping and IID practices of Agile can lead to better software, the fact remains that few executives are willing to write a blank check to the development organization. Most paying customers do not allocate funding and approve the start of work without an agreement on the scope of work and without commitments from the development organization on expected outcomes.

Unfortunately, I've seen my share of projects where the paying customer had already approved funding and the scope of work based on their high-level business needs and funding available without truly understanding the requirements and the effort to develop a working solution. That's usually a recipe for failure. The organization needs to spend some time and effort on scoping the work.

Here again, the practices normally associated with the traditional SDLC model have some advantages. For example, a traditional model makes times for requirements gathering, design assessments, and project planning before starting development. Traditional SDLC practices implement some level of project monitoring and reporting to ensure that the project is tracking with the paying customer's expectations. The traditional SDLC models also enforced some amount of documentation to help protect, support, and sustain the IT assets over their lifetime. Those processes and activities largely fall under the traditional waterfall based SDLC and project management disciplines.

It's important to note that the Agile Manifesto did not specifically state these practices were all wrong, just overdone. In my experience, customers and executive sponsors may expect detailed requirements, architecture, and designs, to be well-defined up front—particularly on large-scale, mission-critical applications. They may also expect to receive periodic and formal project status reports.

Moreover, as much as Agile practitioners abhor excessive documentation, IT employees and consultants come and go over time, and corporate executives must make sure there is adequate documentation to maintain, repair, and extend their critical business applications.

The bottom line is that many organizations retain practices associated with the traditional waterfall SDLC model, and likely will into the future.

Not your Grandfather's Waterfall

It is important to note that we are not talking about your Grandfather's waterfall approach here. For example, there is no reason why a structured, waterfall based SDLC cannot have a Prototype phase or even multiple iterations within a funded product development life cycle. What would separate this approach from other forms of Agile would be more focus on initial analysis and planning activities, additional emphasis on documentation and CM, and more checks and balances against the metrics and deliverables approved by the customer or executive sponsors.

Waterfall Feels Natural

For many executives, customers and managers, waterfall feels more natural. I've had customers who demanded it, after having bad outcomes on Agile-based projects. Under waterfall practices, the paying customers like to be able to authorize a scope of work, allocate budgets, and determine schedule constraints based on the needs of the business. The planned development phases are easy to understand, monitor, and track progress against the plan. Also, the documentation associated with the waterfall approach is appropriate when there are Industry and government regulations that require stringent documentation compliance, such as required under Sarbanes Oxley (SOX).

Under waterfall, the project manager forms the project team, and together they create a detailed plan of how to accomplish the scope of work within the constraints of the project. The executives and managers review and approve the plan, and thereby have a tool to gauge both the progress and success of the efforts going forward. Periodic status reports show progress against the plan.

As the project is kicked off, requirements are gathered and analyzed and drive the technology decisions and design. Once the solution architecture and designs are complete, the developers are free to start creating and then testing the code. Once the tests are complete, demonstrating the product conforms to the original requirements, the IT organization deploys the product into the production environments.

Waterfall, in short, gives the customer a sense of being in control.

Best Fit

Waterfall practices work well only when business and user requirements are steady and not subject to significant change throughout the project. Also, some legacy applications are not easy to adapt to an Agile approach as the software development environments may not support prototyping and short-term iterative development practices. However, even with modern development methods and tools, I've seen waterfall approaches employed when developing large enterprise applications supporting mission-critical business processes that cannot change without a formal rollout across the enterprise.

When developing mission-critical applications, especially those that support the strategic goals of the business and represent significant investments, failure is not an option. Prototyping and IID SDLC approaches can improve the waterfall-based development process. However, the complex architectural and design issues associated with building large-scale enterprise applications often make the detailed requirements analysis and design phases of waterfall a better approach to starting development of those applications.

Documentation and Configuration Management

Agile proponents do not advocate extensive use of documentation. However, from the perspective of an organization's CEOs, these investments must be protected through the documentation and protection of their intellectual property via CM processes and systems. They should understand that developers and contractors come and go, and their knowledge of what components make up the business application or systems, and their understanding of architectural and design decisions they made, go with them without enforcement of appropriate documentation and CM practices.

Configuration management practices and systems track all items developed in support of a product release. The items can include source code and software modules or components, documentation that provides information on requirements, architecture and design decisions, software and systems support instructions, product builds and configurations, test scripts and test results, user guides and training aids. Development organizations must manage and archive all configuration items for every release of a product. The information is critical to debug, sustain and enhance a released product over time. Moreover, the configuration items represent intellectual property (IP) that is expensive to develop, and even more expensive to recreate if it's ever lost.

A company's IP on a business system includes the requirements that drove the features, functions, and designs for the product; the critical business process changes implemented by the various releases, and how architectural, design and component changes associated with each release. Other intellectual property includes data and data management requirements and the methods for restoring, updating, enhancing and maintaining the deployed system.

CM practices evolved to support the technical management needs safeguarding information on hardware material items. The U.S. Department of Defense codified early CM practices in **DOD-STD-480**, *Configuration Control – Engineering Changes, Deviations and Waivers*, in 1968. The DOD updated that standard under **MIL-STD-973**, *Military Standard: Configuration Management* (17 APR 1992). That standard, too, has been superseded by the **EIA649C**, the current *Configuration Management Standard*, revised February 7, 2019.

The plan-driven and heavily engineered nature of traditional SDLC practices made CM practices a mainstay in the early software industry. Nevertheless, agile projects also need to implement rigorous CM practices, to protect the organizations investments in its product releases, for long term supportability and sustainment needs, and to maintain IT based intellectual properties.

How things can go wrong with waterfall

Things can and do go awry, often starting with Project Charter itself. The sponsoring executives likely have gone through a series of strategy sessions to determine their goals and objectives and budgets for the next fiscal year. Where they may get in trouble is making budget, resource and schedule decisions before completing sufficient requirements and design analysis and project planning to test the financial feasibility of the project.

For example, they may complete a thorough business case that includes a cost-benefit and ROI analysis. Such assessments should include estimates for all the peripheral investments they need to make to support production, marketing, sales and delivery of the new and existing product and service offerings. However, their ROI and cost-benefit assessments are only as good as the data they have on development costs.

Without a detailed project plan, created and agreed to by the team that must perform the work, the sponsoring executives do not have a validation of their baseline cost, resource, and schedule estimates. Moreover, without an analysis of business, end user and design requirements, there is just no way the project team can truly understand the total scope of work that is required. In short, the budgets and resources allocated may not have any basis in reality. That can be a real problem.

Let's look at an example of a failed process. In this scenario, we assume that the executives complete their strategic planning sessions and determine that the budgets and resources allocated seem reasonable based on the executive sponsors' ROI and cost-benefit assessments. The next step is to release the project charter and hire or assign the project manager. The project manager helps assemble the project team. The project manager and team, working with other stakeholders, collectively develop the initial project plan.

Note that often a development team is forced to create the initial project plan before business, end user, and design requirements are gathered and analyzed. The lack of detailed requirements leaves a gap in their understanding of the scope of work to be completed, and the risks they must take on. From this author's experience, too many projects get handed over to a project team with only a high-level set of deliverables specified with little or no time allocated by the customer to conduct a requirements phase.

The lack of attention to requirements happens for two reasons. In some cases, the customers and sponsors believe they have documented their requirements sufficiently in the deliverables list. They either don't understand there is not enough information to work with, or they don't have a complete vision for the product yet—and they may expect the developers to know what they want.

That leads to a second problem, which is a failure by the customer or executive sponsors to make their domain experts and end users available during requirements-gathering sessions. For the developers and business analysts to figure out what's required, they need access and time with sponsors, customers, end users, and often other stakeholders. Truthfully, this is a potential problem regardless of the SDLC approach chosen for the project.

If customers, end users, and stakeholders are unwilling to commit their time for requirements gathering and feedback, the likely result is the development of a product that does not meet their needs. Moreover, under traditional waterfall approaches, all the requirements need to be gathered up front, at the start of the project, and all at once, which can be a significant time commitment. This time commitment needs to extend beyond the requirements gathering phase and into the design phase, as the developers need to solicit feedback on the impact of their designs on the business and on end users.

If the designs are wrong, issues are not discovered until the completion of the development phase, forcing time-consuming and costly debugging and rework of the application. To provide a buffer, contingency plans to schedules and budgets help protect against potential delays and budget overruns.

MANAGING PERCEPTIONS

Customers and executive sponsors want to know how well their portfolio investments in software and systems are doing. They want updates showing progress against their budgets, schedules, and deliverables. If a project goes off plan, the project team can expect more attention from management and their customer than they want.

In such situations, you can expect the paying customers and executive sponsors to ask lots of questions, such as: Do we know what's wrong? Can we get the project back on track? If so, what's it going to take to do so? Moreover, they may want to know if they need to pause or kill the project.

Managing perceptions is a critical issue, as there are almost always folks with other priorities and projects in the wings, and the sponsors of those projects—seeing blood in the water—may go after the "failing" project's funds and resources. Alternatively, executive sponsors or customers, sensing a failing project, may decide it is better to pull the plug and cut their losses.

DID YOU KNOW?

A *Risk* is a potential problem that has been identified.

An *Issue* is a risk that has already happened, creating a problem the team must address.

Risks are mitigated; Issues require contingency plans.

MANAGING SCOPE CREEP

If the project team has done a thorough job of developing requirements, created a requirements traceability matrix (RTM), made their design with linkages back to the requirements specified in the RTM, and developed the product in conformance with the RTM/design plans, then testing should go fairly well. However, when the project team takes shortcuts in the requirements phase, the user acceptance testers and field testers may be disappointed with the results, and report features that do not work as they expected as bugs.

True story: I took over a waterfall-based project that was failing. The project team, after a year of development, had not released, in the eyes of the customer, a working product. During user acceptance testing, the testers reported undocumented requirements as bugs—more than 100 in total.

An undocumented requirement is a requirement that didn't make it into the requirements documentation, yet the customer and end users insist it was a missed requirement by the project team. Project and development teams almost never win that battle if there are no requirements documents that have been signed off by the customers.

The previous project manager sided with the customer's end users who were testing the product, and the project immediately started overrunning its initial budget and schedule. The schedule and budget overruns went on for some time as the executive sponsors extended the budget and schedules.

With a list of more than 100 "bugs," really enhancement requests, the project team went into an iterative development mode to accommodate the client's new demands. However, the customer's end users were never satisfied, as each new iteration of the software brought new insights to the end users on how the product could support the organization. The requested enhancements went well beyond the original contract requirements, but the end users realized they could keep asking for new features.

In other words, the PM had given up control of his project, and scope creep took over. After several additional extensions, the executive sponsors were ready to pull the plug, as they saw no end in sight to the schedule and budget overruns.

We had to stop this insanity by disallowing the introduction of any new features and functions, and then work through the client's "bug" list as fast as possible. Our project was fortunate to be extended, we hired additional development and testing resources, and our team delivered a working product into production within the revised schedule and budget constraints. Since the identified defects were missed requirements, the bug list became the basis for developing our product requirements.

We also applied IID practices to rebuild the product with ongoing customer feedback. That alone gave the project a more favorable view among the client's program manager, executive sponsors, and end users.

All this pain did not need to happen in the first place. The original PM failed to realize that the customer and end users did not have a complete vision for the product. They should have spent more time, in the early stages of the project, on requirements analysis and prototyping. The final product was two to three times more expensive than it should have been. Had it not been such a critical requirement, the customer would have, and probably should have, pulled the plug much sooner on the project.

Also, as noted, this is the type of situation where IID makes sense, enabling rapid delivery of working software updates for customer and end user reviews. Without a clear product vision, the customers didn't initially know what they needed in the product.

Bottom line, a high-level deliverables list is not a requirements document, no matter how much your client thinks otherwise. Developing detailed requirements is a must, whether through a traditional requirements and design phase or via prototyping and iterative development cycles—even better, incorporate both strategies.

MANAGE CHANGE

Finally, the other issue with traditional waterfall practices is the impact of change and, as demonstrated in the previous example, the impact of introducing new requirements. Traditional waterfall-based development projects, by their very nature, are chartered to create very specific deliverables within authorized budget, schedule, resource, and quality constraints. The uncertainty of project-based work virtually guarantees some potential for change. Regardless, change on a waterfall-based development project must be rigorously managed.

Risks and issues that crop up can have a negative impact on project constraints. New enhancement or modification requests likewise have a negative impact on the project's constraints. Since waterfall is a linear and sequential process, there is no built-in mechanism to re-evaluate new requirements and priorities in the middle of a development cycle.

The project plan defines the change management processes for the project. The project team submits formal change requests when changes to project constraints are unavoidable. If the proposed change severely impacts the ability of the project team to deliver on budget and schedule, the sponsors are not likely to approve the change. As the paying customers, that is their call to make. In those cases, the needs identified in the change requests may have to be put off until the next release, probably into the next fiscal year.

AGILE

As noted in the previous lesson, Agile is a set of values and principles and not a specific SDLC methodology. Put another way, Agile-based methodologies support the values and principles articulated in the Agile Manifesto. Today, there are numerous SDLC methodologies and tools available that fall into the Agile category, which we have noted in previous sections. In this section, we'll dive deeper into the foundational SDLC concepts that best support Agile practices.

The three foundational SDLC concepts that support Agile practices the most include:

- **Prototyping - a "try before you buy" approach**
- **IID - a "try not to eat the whole elephant at once" approach**
- **RAD - a "let's use some cool tools to get there sooner rather than later" approach**

Prototype Development

As mentioned in the first lesson, the Agile approach chosen for a project is determined based on the characteristics of the project. For example, if the customers and stakeholder know ultimately what high-level capabilities they require, but don't have a vision or starting point for how the product might implement those capabilities, then a prototype model is probably a really good idea. Prototyping also works well to demonstrate the capabilities of COTS products, or to demonstrate concepts or approaches to solving a problem.

PROTOTYPES ARE A THROWAWAY EFFORT

Executive sponsors and customers need to be aware that a prototype is not a fully functioning product, and in fact may not create any useful artifacts. The main objective of prototyping is to try out various ideas to see what works and what doesn't. If some of the artifacts are useable, that's a bonus. If not, the effort may lead to new ideas to test out. At the very least, the prototyping effort may demonstrate that continued investments on a particular development strategy or product concept are not worth pursuing.

Certified project managers know the definition of "sunk costs" (money and time spent on unproductive work and failing deliverables that cannot be recovered). When mistakes happen, learn from them, but move on. A prototype as a throwaway effort is an example of that. The project has born the costs, usable or not, and the project team moves on accordingly with the knowledge gained from the effort.

Letting go is not a bad thing, assuming the project team plans for it and the sponsors and customers have bought into the approach. In short, the customer or sponsor is paying for a learning experience that is cheaper and easier to bear than making a wrong decision and having a failed project.

IID

In the IID model, one can imagine the waterfall model compressed and repeated across shorter time intervals. The iteration part of the model is a cycle of repeated steps—such as requirements gathering and analysis, design updates, development, testing, and customer feedback. Each cycle of process iteration incrementally adds to the overall functionality of the product. This Agile approach is particularly useful when customers want to deploy releases of useful functionality bit by bit, rather than waiting until the entire solution is designed, developed, tested, and ready for deployment. It is also a useful method to help customers and end users work through concepts of how the product needs to evolve to support their business practices or other needs.

TRIAL AND ERROR

Software applications are unique, as far as products go, in that an application can change the very nature of the work it supports. Often, customers and end users do not have a complete vision for a product but can gain important insights through trial and error over multiple development iterations. In such situations, the evolution of a workable solution comes from the layering and integration of new and improved capabilities over time.

ARCHITECTURE AND DESIGN ENHANCEMENTS

With IID, poor architecture, design choices, and bad code can be found and fixed early, instead of being embedded or deeply entrenched in a final release. On the other hand, making early design decisions can come at a cost if they are wrong. An insufficient understanding of the total solution requirements, at the start of a project, can lead to poor designs that create more work to fix after the application has evolved through multiple iterations and product releases.

RAD

RAD is a tools-centric approach to rapid development, and often includes GUI development tools and a library of reusable software components. RAD methodologies have a focus on prototyping, iterative development, and timeboxed development iterations. RAD tools typically employ object-oriented programming (OOP) software design and development practices to develop and reuse common software components.

Proponents of RAD believe faster and more efficient development processes and tools lead to better, higher quality software. While RAD tools still exist—with a focus on visual development techniques over software coding, the modern practice has trended toward the use of IDEs) to improve reuse and efficiencies in software development.

As with Agile, in its modern contextual use, RAD is more of a conceptual approach than a specific development methodology. Also, like Agile, RAD approaches employ rapid prototyping and iterative development practices. RAD is differentiated in the use of tools that facilitate rapid prototyping and frequent informal communications with customers and end users. The underlying tools and technology drive variations in the capabilities of a particular RAD approach.

With an emphasis on improving the speed of development, the focus is on evolving the product through rapid prototyping. When developing software applications, design issues and bugs are a given. Under RAD concepts, such issues are addressed in later releases, as opposed to getting bogged down debugging a product in the early stages (iterations) of development.

RAD TOOLS

OO software development tools and visually oriented GUI builders help simplify and speed up the development process. These capabilities make it easier to involve customers and end users in the software designs from the beginning.

MDD tools fall into the category of RAD tools. MDD tools implement diagrammatic or other visual approaches to describe how a proposed software system should work. When a design is ready to build, the MDD tools automate the process of generating code. Some MDD tools also simulate the working of the code before building it, and also automate software testing. A RAD environment may include one or more of the following capabilities and tools:

- **Requirements Documentation**—such as the use-case model capabilities
- implemented in UML tools
- **Prototyping**—Can include GUI builders, **4th** generation programming languages, Web application frameworks, software factories, simulation software, and other approaches.
- **Computer-aided software engineering (CASE) tools—CASE tools support** specific tasks across the software development life cycle. The idea behind CASE tools was to apply computer-aided design (CAD) concepts—used to design mechanical systems and hardware, to software development. The CASE approach provides software tools to model business processes, data, and objects. OO concepts drove the evolution of CASE tools. Over time, the OO methods were consolidated and encapsulated in the standards maintained today by the Object Management Group.[82] The CASE term was popular in the 1980s through 1990s but now is considered rather dated.
- **IDE**—an integrated suite of tools that support coding of the application, such as source code editors, build automation tools, code debuggers, **Intelligent** code completion features, compilers, interpreters, software version control systems, GUI Builders, and OO development tools.

EMPHASIS ON SPEED

RAD borrows from the development concepts of prototyping and IID with an emphasis on using software development tools to expedite the software development process. An advantage of RAD is that customers and end users can see their ideas instantiated in the product more quickly.

As with other prototyping practices, RAD-based GUI builders enable rapid trial and error of different user interface concepts. The benefits come from reductions in time, effort, and costs associated with developing RAD prototypes. Modern OO programming languages offer prebuilt modules that can be reused across any number of applications and enable rapid construction of software components, helping speed up development of the logical and data flow elements within an application. These modules act as building blocks for assembly and incremental testing. In short, RAD implements a form of IID, only quicker (e.g., IID on steroids).

HOW THINGS CAN GO WRONG WITH AGILE

If customers and users think requirements gathering sessions under waterfall are overly burdensome, then chances are they have not experienced an Agile-based development project. Most software developers are not domain experts in the areas they support when building an application. As a result, they often have very limited knowledge of the business processes the product supports or the desired product capabilities that their customers need.

Moreover, the developers cannot make product design decisions on behalf of their clients when they do not know the impact those decisions have on customers, stakeholders, and end users. So, while Agile-based practices accommodate frequent customer reviews, it is all to no avail if the customers and end users do not take the time to participate in incremental reviews and provide their feedback. Without customer participation, RAD tools and IDE suites just let developers create useless software more quickly.

Executives [should] have an idea of the system's capabilities they need to operate their business and enter new markets. They also [should] know how much they can afford to spend, and deadlines that must be met. Agile practices can be problematic if the developers lose sight of these issues, and are not sensitive to the client's scope, budget, and schedule constraints.

For example, if the developers get off into the weeds supporting the implementation of the end users' requests for bells and whistles, but not anticipated by the paying customer, the schedule and budget constraints can be quickly blown through. In such cases, the executives might view the project as failing. Product managers must also stay attuned to the needs of the business when making backlog priority assessments. Blown budgets and schedules have an impact on the organization's bottom line.

Finally, the development team must provide basic reporting to show progress against specific project metrics established by the executive sponsors or customers. Most Agile reporting methods are focused on assessing the efficiencies, capabilities, and progress of the development team. Those assessments must be mapped back to the project's constraints of scope, quality, schedules, and budgets.

Again, without adequate reporting, the executives might view the project as failing, which could lead to significant management attention in a best-case scenario or lead to a shutdown of the project in a worst-case scenario.

HYBRID SDLC MODELS

In the introduction to this chapter, I noted the 2019 CollabNet | VersionOne polling results found that more than 48% of the respondents indicated fewer than 50% of their projects are following agile practices.[83] I've also noted that the need to build and maintain a repository of best practices, around the tools and technologies employed by the organization, exists even under agile practices. Given that both these conditions exist, most organizations need a way to manage information on both traditional and agile processes and best practices.

The implementation of traditional SDLC processes with Agile-based processes within an integrated SDLC Framework is referred to generically as a hybrid SDLC model. Under modern hybrid approaches, the project team implements iterative development cycles and multiple product releases during the approved schedule while preserving some of the structured processes, engineering disciplines, and documentation required under the waterfall approach.

Modern hybrid approaches typically maintain the pre-development phases and processes for detailed requirements gathering and analysis, project planning, architecture, and system design. However, at the development stage, the hybrid model may offer a choice of two paths: one to continue along the traditional linear sequential flow of work; or, two, to conduct development and testing of individual releases following agile practices.

That's not the only way a hybrid model may work. For example, a hybrid SDLC may look more like a toolkit wherein any number of systems development approaches are documented and made available to the organization's software and systems development teams. The Disciplined Agile Toolkit is an example of this approach.

I'm not going to spend much time talking about hybrid models here, as the subject is covered more extensively in my book titled *Frameworks for Agility*. While different hybrid approaches are discussed, my *Frameworks for Agility* book promotes a hybrid SDLC framework to manage engineering practices in the traditional phase base approach, while also managing agile processes that can be integrated with the engineering practices at the program or project level. Also, in chapter 12 of this book, I discuss how the project manager can integrate multiple SDLC and PM processes, tailoring their project's activities and schedules to meet their unique needs.

A View of Sustainability

Large enterprise applications can develop over time with Agile practices. In my *Frameworks for Agility* book, I discuss the concepts of evolutionary architectures and designs. Rather than the "*Big Upfront Design*" approach favored under the traditional model, agile projects need their architectures and designs to be flexible and evolve incrementally to support changing requirements and feedback.

This is not to say that architecture and design processes are less important in agile projects. If the developers do not apply the same type of rigor with regard to establishing a sustainable architecture and design, and they do not provide adequate documentation of the various software builds, associated requirements, interfaces, and integration mechanisms created in the numerous increments, the long-term sustainability of the application is compromised.

DESIGN, BUILD & TEST FOR PERFORMANCE

Invariably, performance testing is one of the last steps in the development cycle, as the application must be built before it can be tested under load and put into a production test environment that mimics the actual production environment. There is nothing that is worse than getting a product through the development cycle, past acceptance testing, and presumable ready for release, only to find out there are significant performance issues.

Application loads, size of data, complexity of the application's logic, competition for computing resources in virtual environments, and integration with other systems can all create performance issues. To the greatest extent possible, such factors must be considered from the very beginning and steps taken to create workable architectures and designs.

Performance testing should not be held off until just before the product is ready for release. Ideally, such tests are part of the ongoing test strategy for each iterative release, whether shipped or not. If the product can't undergo ongoing performance testing, the release is at risk.

For example, I have seen development teams back out features prior to a release when performance testing was scheduled too late to discover and resolve the underlying performance problems. So, always make sure there is ample time in the schedule to address any issues after the final performance test is completed, prior to a production release.

CHAPTER SUMMARY

This chapter has presented an argument that a mature SDLC should include the best of both traditional waterfall and Agile-based development practices. Such an SDLC approach is sometimes referred to as a hybrid SDLC model or hybrid SDLC framework.

While the traditional linear-sequential waterfall approach to development has major faults, there remain practical applications of some aspects of the traditional model. For example:

- Detailed requirements and design analysis before starting development reduce the chance of costly and time-consuming redesigns and rework as the product evolves.
- Enforcement of standards in technology and enterprise architecture help ensure long term and cost-effective supportability and sustainability of developed software products and systems.
- Documentation and maturation of development and engineering practices helps the development organization improve the quality and consistency of their output.
- CM practices help protect the organization's IT investments.
- Product deployment processes facilitate information collection and transfer to the IT **Operations** groups who are responsible for the product's release, including installation and configuration activities, plus ongoing support and sustainment activities

Agile approaches evolved to address the shortfalls of the linear-sequential development approach. The shortfalls of the linear-sequential development process include:

- Lengthy development cycles—often over an annual budget cycle or longer
- New functionality offered infrequently
- Code defects that are difficult and time-consuming to debug when waiting to finish all coding

- **Little to no customer input during development**
- **Inflexibility to support changing requirements or an evolved understanding of requirements**
- **Overly burdensome reporting and documentation requirements**

We learned that Agile is not so much a specific development approach as it is a set of values and principles driving modern software and systems development practices. Modern software development tools have made Agile values and principles increasingly practical. Today, the term Agile encompasses a wide variety of development methods that have diverged from the traditional waterfall patterns.

In general, a modern corporate SDLC allows project teams to mix and match development approaches to suit the needs of their project. Some organizations have two tracks within the standard corporate SDLC—one for categorizing and managing their engineering and best SDLC practices under traditional development life cycle phases, and another to provide guidance on Agile-based methodologies. Other SDLC frameworks, such as disciplined Agile, support the deployment of a multitude of SDLC practices, with guidance on how and when to employ specific approaches on a project.

Project and product managers should be aware that the product development practitioners, who previously used nonstandard methods and tools, especially when they have proficiency in them, may form biases for a software or systems development approach. Use of nonstandard processes, methods, and tools present a danger to the success of the project, and the long-term support and sustainability of a product.

Finally, management expectations and the level of customer and end user interactivity are factors to consider when choosing a development approach within the SDLC framework. The achievements of the project team must be communicated in terms of the objectives that made funding possible. Moreover, while Agile disciplines are wonderful in providing opportunities for customer input, the customers must commit their time and other resources to the periodic reviews to obtain the benefits.

The next chapter provides a discussion on the business drivers that change the way the organization must conduct its business, and how life cycle development processes apply across the enablers of the business, not just IT. Moreover, when developing business supporting applications, the SDLC processes must support the organization's other life cycle change processes.

06.
Turning Strategy into Action

This chapter will be useful for readers who want to understand how corporate strategies and strategic plans help align the organization to address external factors impacting the business, the use of portfolio, program, and project management to execute those strategies, and execution options available to affected IT departments..

CHAPTER OBJECTIVES

At the end of this chapter, the student should be able to discuss the following:

- t are organizational/business change drivers?
- What is the purpose of Strategic planning?
- What is Portfolio management?
- What are the differences between program and project management?
- Discuss various approaches to estimating development costs during initial planning phases.
- Explain the purpose of outsourcing work and various contract options.
- Discuss the dangers of shortchanging SDLC processes.

AUTHOR'S INTRODUCTION

This chapter begins with a discussion of typical business drivers that cause an organization to change, and how executives initiate change activities in support of the mission of their business through their strategic plans. It's likely you've never seen a strategic plan handed down to employees instead of a project plan. That is not a sound idea as the strategy is at too high of a level to plan and execute work.

At the conclusion of strategic planning activities, the executives know they need some parts of the business to change and how much they can afford to spend. There's more to the effort than the last sentence implies. For example, the executives have evaluated the mission and vision of the organization and conducted some analysis on market opportunities and the competition. The executives have some idea of the deliverables required to support the mission of the business, current vision, and associated change requirements. They should know what types of products and services their customers need, who their target market customers are and what they can afford to pay, and who the competition is. They should have an idea of how much they can spend on the change effort to stay in budget and remain profitable. Moreover, the executives should construct a roadmap that outlines the expected schedule of funded deliverables.

Okay, that sounds like we know a lot. However, now what? There are still many questions we don't have answers to. For example:

- **How do we align the efforts of the impacted organization, both inside and outside the company, to achieve the objectives and goals outlined in the strategic business plan?**
- **Do we know how much all the change initiatives will cost?**
- **If the costs are greater than the budget, do we know where we can cut back?**
- **Do we know which business processes must change and how to support the new business strategy?**
- **Do we know yet if our systems can support the new business strategies?**

There is so much work left to do before we can answer these types of questions.

While this book is about SDLCs, we need to understand how strategic planning leads to planning activities that eventually impact the organization's IT departments. Specifically, we need to understand how portfolio, program, and project planning activities help the organization translate their strategic plans into actionable plans.

The challenge facing executives is how to plan appropriate levels of funding and resources to support their business strategy. That's a hard thing to do if some degree of tactical planning is not going on in parallel with strategic planning. Fortunately, there are proven approaches to conducting both strategic and tactical planning in parallel.

A typical approach is to break the tactical planning work into three buckets of work: portfolio management, program management, and project management. Portfolio, program, and project management are separate disciplines with unique processes, although program and project management processes are similar. All three of these management techniques overlap and integrate via activity dependencies.

In the course of my career, I've seen only a handful of companies do this type of integrated planning very well. As with the implementation of SDLC processes, the organization must invest time and resources to get these activities right. Moreover, the executives must be fully invested in these efforts, communicating their support and monitoring results, to ensure organizational buy-in and follow through.

Planning in larger organizations with multiple divisions is more complex, and the process can be much more laborious and time-consuming. In such cases, each division may have different products, different customers, and separate marketing, sales, product development, and delivery organizations. If the organization's executives send out their corporate strategies at the start of the fiscal year, can they expect the organization to turn on a dime and make the adjustments to meet the strategic goals and objectives for the new fiscal year? Again, of course not. The organizations need to back up the efforts with enough lead time to align and integrate their budgeting and planning activities.

So, this chapter discusses how organizations often get it wrong, plus methods for getting it right, with regard to driving strategic plans into actionable plans. In this discussion, I'll explain the differences between portfolio, program, and project management. I'll discuss IT development options and risks when the strategic plans impact IT investment decisions. I'll also discuss some contract options and the pros and cons of each. I'll discuss the dangers of shortchanging the life cycle processes. Finally, the chapter concludes with a discussion on the potential impact outsourcing development work has on the IT organization's ability to control the delivery and future sustainability of a software-based application or system, and how to minimize those issues.

WHY CHANGE

Organizational change is inevitable, as all enterprises need to evolve to stay in business and relevant. Businesses that aren't open to change will be undercut by their competitors. Nonprofits that don't change may provide the wrong solutions or fail to attract donors. Government agencies that fail to change may deliver the wrong services or deliver services that are inefficient and costlier than they need to be.

In Chapter 1, in the ECM, I noted that life cycle processes facilitate change across the enablers of the business. The enablers of the business include anything that supports the mission of the business, but nothing else. That is the concept behind lean practices—the elimination of anything that does not add value to the customer.

If our organization is delivering products customers don't want, that is waste, and we need to change. Organizations also have waste when they can't deliver a product at a competitive price or with competitive features and functions. At some level, all forms of waste result from ineffectual processes implemented within the organization.

In competitive or changing business environments, target market customers and business opportunities constantly change. Affected business processes are almost always ripe for change. As discussed in the ECM section in Chapter 1, technology can change—which can in turn change necessary skills and processes. The information we need to manage our business can change.

As IT specialists, we care because the change in any of the enablers of the business (e.g., people, process, information, technology, customers and products) can impact the information technology needs of the organization. Business change requirements drive changes in IT development and operations SDLC processes encompassing infrastructure, networks, software applications, and integrated systems.

Business Drivers

External factors often drive the need for organizational change. For example, the commercial enterprise is under constant competitive pressures from other companies within their industry that attempt to take their market share, revenues, and profitability by introducing new products that may have different sets of features, better pricing, or better quality. The business entity may need to respond in kind and seek out a new niche within their target market or new markets altogether to sustain their growth and profitability.

Government agencies are influenced by voters, at least indirectly by the parties that control the executive and legislative branches. New cabinet members often come in and influence the types of citizen services offered by the department, so that they are in alignment with their party's objectives. However, they may also be directly influenced by the press and citizens when there is a perception that the organization is not serving the best interests of citizens, special interests, and business.

Nonprofit agencies are affected by changes in the economy, which can impact their donors. Changes in political parties, policies, and government taxes can impact the nonprofit, forcing them to adjust their business model and processes. It is critical to business operations to have up-to-date information on their donors, volunteers, and recipients of services. The needs of these individuals can all change over time, driving new business decisions. Also, technology and the best methods for providing benefits may change over time.

BUSINESS STRATEGY

Business strategies are developed by the chief executives to ensure organizational alignment with the mission of the business. A business strategy succinctly states what the organization's business is, the customers they serve, the corporate goals and objectives set by the executives, how the organization is expected to achieve the goals and objectives, and what "good" looks like when achieving the goals of the strategy.

At the heart of any business strategy is providing continued support toward the corporate vision for the business. A vision statement provides an outward look at what the business should be in some ideal future state. The vision statement sets the direction and aspirational goals for the organization.

The vision statement for a business doesn't usually change unless there has been a major shift in the industry that requires a change in the focus or purpose of the business to sustain the viability of the organization. What can change are the needs of the customers or the market, and how the organization is aligned and operated to deliver the right set of products and the right prices and respond to current market conditions. That is the purpose of the mission statement.

A mission statement succinctly but clearly states why the business or organization exists, who the organizations support (e.g., customers), and the benefits provided by the organization to its customers. In short, the mission statement communicates the purpose of the business or organization. The mission statement also represents the current state of the business or organization, in contrast to the vision statement's longer-term view of the business.

For example, the truck division within an automobile manufacturer doesn't just build and sell pickup trucks. Instead, they may sell competitively priced 4-wheel drive pickups that are tough enough to explore the trails of the Rocky Mountains while tame enough to be comfortable on the freeways driving home. Admittedly, that may sound more like a marketing pitch, but it covers the basics of what we build, who we build the products for, and why our customers should care.

This should be obvious: if the organization doesn't know who their customers are, or why the customers need them, they probably don't have a viable business proposition.

Strategic Planning

At the beginning of each fiscal cycle, an organization's chief executives typically get together to discuss the needs of the organization over the next fiscal year. They may start with a review of their vision statement to make sure they understand and agree on what the organization's purpose is, and to review whether the policies, procedures, projects, programs, and business operations they have in place are correctly aligned to support the vision of the business.

The executives should also review how well they achieved their plans from the previous year, and use those lessons learned to build a better plan for the next fiscal year. In short, the executives must question whether the enterprise was doing everything it could to provide the right set of products and services for the right set of customers while operating within their budget, time, and resource constraints.

As part of the strategic planning exercise, the executives review information on current market conditions, emergent customer needs and new opportunities, and their ability to achieve their goals from the previous year. Once this initial assessment is complete, the executive management team must turn their attention to developing the mission statement for the upcoming year. In effect, the mission statement provides an opportunity for the executives to communicate necessary course corrections for the organization.

The objective of a business strategy is to ensure the alignment of organizational resources, investments, and processes in support of the mission of the business. As part of that alignment, the executives must agree and approve a portfolio of investment decisions that support the business strategy. Finally, they must set goals and measurements that demonstrate progress against the strategy. Ideally, programs and projects are set up to monitor, control, and report on the progress of the organization's strategic investments. Most importantly, as part of the strategic planning process, they must ensure that organizational resources are appropriately aligned to support the mission of the business.

Strategic Goals and Objectives

If the entity is a commercial business, development of the strategic plan is all about ensuring competitive advantage, growth, and profitability. If the organization is a government agency, their strategic planning has a focus on ensuring the agency is efficiently providing useful citizen services mandated within the scope of their commission. Likewise, a nonprofit organization may develop a strategic plan to specify the work they want to accomplish in the upcoming year and document their anticipated funding requirements and targeted sources of income or contributions.

Course Corrections

Things do not always go as planned. When an unexpected problem arises, the organization must make course corrections during the fiscal year. Some organizations may set aside funds for midyear corrections plus provide a funding source for projects that support emergent business needs. Also, well-managed organizations implement periodic reviews of existing programs and projects to ensure that the goals and objectives are met within the budgets and time allocated, following the strategic plan.

Milestone reviews or *phase gate reviews* are the names given to these periodic reviews. If a project is failing to achieve its scope or exceeding budgets and schedules, the executive team may choose to grant more time and resources to address the problems. On the other hand, they may choose to shut the project down and reallocate their funds to other critical needs within the organization. Priority changes within the organization can have the same effect. Funds can be taken from one program or project and applied to another.

Metrics—what "Good" Looks Like

Defining a set of strategic goals and objectives is of little value if the executives do not also determine what good looks like in the form of quantitative and qualitative metrics. The quantitative metrics may include things like cost of production, production throughput, ROI figures, product pricing, number of customers, total sales revenues, and profit margins. Qualitative metrics usually take the form of written or spoken descriptions and are often used in marketing and sales where the company needs to articulate its value proposition and other messages to prospective customers.

Qualitative metrics are also useful in describing expected outcomes for the training of end users on new products and services, or when the organization needs to explain its position on critical issues to stakeholders, government agencies, or customers.

Change is Inevitable

In summary, periodic organizational changes are necessary to help ensure the business remains aligned with the mission of the business. A key part of strategic planning is to determine what products and services must be provided by the organization to meet the needs of clients or constituencies. When the products and services have to change, investments are required to make those changes, and executives establish funding and timelines based on expected ROI.

Executives establish goals and objectives with specific measurements so that everyone in the organization knows the expectations when executing work against the strategic plan. Programs and projects are initiated to implement organizational change to achieve the goals and objectives established by the strategic plans and directives.

A critical challenge is detailed project planning from the very first time the organization finally has the project team and other SMEs working to determine the sufficiency of resources, budget, and time to meet scope and quality constraints. The executives must be willing to reassess their investment decisions if the approved project constraints are insufficient. Also, the organization must make course corrections as needed when new information becomes available.

STRATEGIC PLANNING DRIVES PORTFOLIO, PROGRAM, AND PROJECT PLANNING

When developing a strategic plan, chief executives determine and agree on the goals and objectives that are necessary to support the strategic plan. They also determine their budgets and how much money they are willing to invest in the various programs, projects, and operations of the business.

For example, when making budget allocation decisions, a commercial company may determine how funds are allocated based on expected ROI or discounted cash flows (DCF). Similarly, a government or nonprofit organization may employ a social return on investment (SROI) analysis to make their investment decisions. Where an ROI is a profit-centric measurement, an SROI is a principles- and values-based impact assessment.

At this level of strategic planning, the budgets and resource commitments are at best very high level and not completely vetted by those who are responsible for managing and performing the work. The goal at this point is to ensure that the organization is working on the right set of programs, projects, and operations that support the strategic plan goals and objectives set by the chief executives.

In Chapter 1, we noted that the scope of work, budgets, and resources authorized in a project charter could be out of line with reality. It's a mistake to think the type of analysis conducted during strategic planning provides a rigorous assessment of the organization's ability to deliver within the desired constraints of scope, time and budget. The organization has more work to do.

At some point, there must be a more detailed analysis to understand the total scope of work that is required to provide the deliverables expected by the sponsors or chief executives and thereby derive better budget, schedule, and resource estimates. The sooner that work is completed, the better. The detailed analysis should occur in parallel with strategic planning to minimize the chance of devoting too much time and resources on an unfeasible plan.

The concept phase of work within my book titled *Frameworks for Agility* contains several methods to evaluate the feasibility of a program or project, such as:

- **CONOPS investigations**
- **Business plans**
- **Value proposition**
- **Feasibility assessments**
- **Cost benefit analysis**

Figure 17- Intersection of Strategic Planning, SDLC - Product Conceptualization, and Program and Project Management, provides a graphical depiction of the intersection of these feasibility assessment activities. The concept phase serves as a boundary line between strategic planning and the beginning of the SDLC. These are the types of analyses that help the organization determine the feasibility of proposed investments. However, I would also argue that some level of requirements analysis and design work is necessary to get a realistic handle on the scope of work, resources, budgets, and schedules necessary to successfully deliver on a portfolio investment initiative.

In the discipline of project management, project portfolio management is an approach to select, improve, and provide oversight on an organization's project-related investments. In effect, strategic planning, portfolio management, product conceptualizing, and project management all intersect in this boundary area. Project portfolio investment decisions are decided based on the assessments completed in the activities defined in the concept phase of work, organized potentially for efficiency as programs, and managed individually at the project level.

Figure 15. Business Strategy drives Strategic Planning, Product Conceptualization, and Program Management

Portfolio Management

The business strategy drives funding decisions across the organization, with some of the funding allotted to the IT organizations. Portfolio management in the context of this book on systems development processes includes organizational processes to facilitate decision making about which corporate IT investments best support the organization's ability to achieve its strategic goals and objectives. Also, under the umbrella of portfolio management, alternative investment strategies are explored and analyzed to minimize organizational risks.

Some of the funds allocated to IT go toward supporting operations, with the rest going to funding IT programs and projects. New programs and processes are established to manage each of the investments allocated to change-oriented IT initiatives.

The concepts behind project portfolio management (PPM) and enterprise project portfolio management (EPPM) go beyond portfolio decision making to include full project life cycle management. PPM and EPPM processes span project and program need assessments and alignment with the corporate strategy, cost-benefit analysis, project resource planning and management, management of project dependencies, project change control, financial management, and project reporting and communications.

Program and Project Management

Fundamentally, a program is a collection of projects that are usefully related and managed together to achieve the goals of the organization. The relationship may be the types of products managed under the umbrella of the program, types of products produced within a division, or some other relationship where collective management supports the organization's goals. An objective of program management is to utilize corporate resources in the management of multiple related projects most efficiently.

Program management offices (PMOs) are sometimes established to manage the portfolio of projects allocated to a program. In general, there are three types of PMOs:

- **Supportive PMO provides expertise on the use of policies, methodologies, tools, templates, forms, best practices, and reporting documentation to project teams**
- **Controlling PMO enforces and monitors use of specific governance policies, methodologies, processes, tools, templates, and reporting documentation used by the project teams**
- **Directive PMO provides resources for managing the projects, as well as the enforcement of PM governance policies and practices provided in the controlling PMO model**

Project-oriented work is differentiated from operations-oriented work by the degree of uncertainty, complexity, and uniqueness of the work. Projects also have a specific beginning and end, starting with approvals to proceed and ending upon the achievement of the project's goals. Because of the uniqueness of work, the details of how to deliver project-oriented work are initially defined in the project plan, but the project plan must be updated throughout the project as new information becomes available.

With these characteristics in mind, it should be apparent that the work associated with most change-oriented initiatives falls into the project category. Therefore, work associated with portfolio investments that change the way the organization does its business is most logically managed as a project.

WHERE ORGANIZATIONS GET IT WRONG

Organizations get it wrong when they confuse the list of deliverables specified in a project charter or request for proposal (RFP) as being a complete understanding of requirements. Likewise, an ROI analysis does not have any real meaning with regard to understanding the total scope of work that is required to create those deliverables. Therefore, it is impossible to determine project budgets, resources, and schedules accurately at the early stages of strategic planning or even during the SDLC phase of product conceptualization.

Project planning helps overcome the deficiencies in strategic planning by mapping project deliverables to the organization's standard SDLC and project management processes. The project manager also needs access to project team members and qualified SMEs who can scope the work and provide accurate time and resource estimates for the anticipated work. Those experts also help assess the project's risks and develop risk mitigation strategies and contingency plans. In other words, the project planning exercise does the work that gets the organization closer to an accurate determination of risks, budgets, resources, and schedules. However, even after constructing a relatively detailed initial project plan, the project's estimates are incomplete. The project plan is actionable but may lack critical information necessary to scope the work accurately.

If the project is developing a simple and relatively small information-oriented web application, design and architecture can be figured out on the fly during the iterative development cycles, and there is less risk of the scope of the project changing significantly. However, larger enterprise applications supporting critical business processes require much more rigor to scope accurately. The estimates provided in the initial project plan are at risk until both the functional and systems requirements analyses are complete. Functional requirements document business and user needs, while the systems requirements include the architecture and designs necessary to implement the functional requirements. The business executives may not understand how much risk remains when funding and approving a large-scale development project before completing the initial requirements and design assessments.

HOW TO GET IT RIGHT

So how is a program or project ever going to be successful when the executives or sponsors already think they have done the work to approve a project during their strategic planning efforts? They know at a high level what they want and how much they can afford to spend. The problem is, at this point of strategic planning, the value of the work planned and authorized has no direct correlation with the actual costs, resources, and time required to create the deliverables. There is no easy answer or shortcut to address this problem of getting enough information to scope the work and identify and mitigate the risks of starting a new project. Fortunately, there are a couple of options available to help the organization get more useful information during the strategic planning phase. For example:

- **External bid**—The organization can submit a request for information (RFI) or RFP to solicit opinions from external industry experts and internal SMEs on expected time, resource and budget estimates before allocating funds for the program or project. However, it's incumbent upon the customer to ensure that they provide sufficient information on requirements to bid the project accurately.

- **Product conceptualization**—The executives may obtain additional requirements information by conducting the CONOPS investigation, business requirements analysis, and product conceptualization activities as a separate project or set of projects.

  A CONOPS investigation is a high-level capability needs assessment study, based on input from prospective users. The goal of a CONOPS investigation is to identify the functional requirements of a proposed system necessary to support the mission of the business.

  A business requirements analysis is a more rigorous investment in requirements gathering and analysis activities. In the traditional waterfall model, this is a separate phase within the SDLC. I've led multiple projects where the entire scope of work was limited to a business requirements analysis.

  The executives should request feasibility studies, business cases, and cost benefit analyses, to better understand the potential value of their planned investments. The processes, methods, and tools to conduct these assessments lie in the concept phase of work within the SDLC. However, note that without a CONOPS investigation and business requirements analysis, the executives lack accurate assessments on the scope of work.

  Upon completion of these types of studies, the sponsoring executives should have a better understanding of what high-level features and functions are required in the solution to support the mission of the enterprise. Moreover, these assessments help evaluate the market opportunity, evaluate alternative approaches and recommendations, and document the benefits of the proposed solution relative to costs.

- **Contingency funds and plans**—There are two parts to contingency planning. From a risk management perspective, contingency plans define the approach to address identified risks that become issues. From a financial point of view, contingency funds may be set aside to cover unanticipated work or resource shortfalls.

  In risk management, mitigation strategies define approaches to prevent or reduce the impact of identified risks. However, stuff happens, and risks and issues occur in virtually every project. Be prepared through proper risk mitigation and contingency planning.

  Instead of funding an initial requirements analysis project, the organization can develop contingency plans to make additional funds and resources available, if necessary, after completion of the requirements, analysis, and design activities. This option provides some flexibility if the proposed solution turns out to be more complex or expensive than originally planned.

- **Defer development costs—The organization can defer final decisions on priorities and allocation of funds and resources to development activities, until completion of the requirements analysis and systems design. In this option, executives defer making their final budget and resource allocation decisions until they have a better understanding of the total scope of work, plus the risks anticipated with the project.**

MANAGING RISKS

All four presented options have risks associated with them. In other words, managing the risks associated with uncertainty presents a separate set of risks. In the following subsections, we take each option in turn to discuss the risks.

Risks with external bids

Unless the customer has completed a rigorous requirements gathering and analysis activity and published their results, those vendors who respond to the client's RFI or RFP lack the details necessary to understand the complete scope of work necessary to produce a viable solution. Until requirements are fully refined, which drives architecture and systems designs, there is a risk that the proposed solution does not address the customer's needs. As a result, the project may be under-scoped.

Risks with initial product concept assessments

CONOPS investigations, business cases, feasibility studies and cost-benefit analyses are focused on capability needs assessments, benefits, and value to the organization. Typically, such high-level business analysis answers: what do we need to invest in, why, and what's the value to the organization? The business requirements analysis drives the understanding of requirements to a level where detailed project planning is possible.

While the product concept and business case assessments are valuable, there is still great risk in relying on those estimates to determine the scope of work. The business-oriented analysis does not provide detailed business and end user requirements, architecture and design assessments, and detailed project planning that leads to workable solution designs and accurate budget, resource, and time estimates.

The reality in every organization is that there are limited resources available to execute the desired business strategy. What good does it do to approve a project, based on its business value, only to find out months into the execution phase that the scope of work was underestimated?

Changes in a project's scope of work, budget, and schedules, if known up front, would impact portfolio decision making. Therefore, it makes sense to conduct strategic planning, portfolio management, product conceptualization, and high-level requirements and design activities in parallel before making commitments to the final portfolio of projects. Executing such a strategy takes organizational discipline and management commitments.

Risks with contingency plans

With this option, there is recognition by management that the risks and scope of work may not be fully detailed, and that the budgets may therefore be insufficient to cover all unknown contingencies. Contingency plans provide a limited buffer against scope changes, schedule delays, and budget overruns. However, a contingency plan or contingency funds cannot be an excuse for poor project scoping and planning.

It is difficult to assess the level of contingency required when the organization hasn't done the work to fully assess business and end user requirements, let alone architectural and design requirements. The more complexity and unknowns on a project, the greater the risk. The greater the risk, the less likely a contingency plan is a viable solution to fixing a broken project. More accurate scoping of work at the beginning of a project reduces the number of potential risks, and also enables more informed mitigation strategies and contingency plans.

Contingency planning is always required. However, contingency funding tends to be better suited for projects that are similar in scope and complexity to others completed by the organization and project team members. That's not to say the organization shouldn't set aside some funds for unexpected contingencies. That's simply prudent financial planning and risk management.

Risks with deferring development costs

Deferring development decisions is a practical solution when the organization's executives have the patience and ability to initially fund and complete detailed product concept analysis, project planning, requirements gathering, and systems analysis activities before making go/no-go decisions on funding the rest of the project. This strategy works best when the organization has the discipline to perform such planning and analysis in advance of the development budget cycle, or in advance of the approaching fiscal year. But not every organization has the discipline and management commitment required to implement this level of coordination and work on an annual basis.

The best way to provide timely information to the strategic planners is to start the project planning and requirements analysis processes as an initial project, perhaps a few months before the start of the next fiscal year. Alternatively, the organization may defer making final investment decisions on development projects until after the start of the fiscal year, as the requirements and scoping data become available. Either approach may include the option to send out RFIs with enough time to get responses back and reviewed before making final investment decisions.

This option may seem counter to Agile practices, but it doesn't have to be. Integrated requirements gathering, architecture and design analysis, and prototyping can be the focus of work on the initial development iterations. Moreover, executive sponsors and customers may prefer some level of requirements analysis and design assessments to be completed before they feel comfortable making long-term financial and resource commitments. Again, that's just being sensible in the allocation and use of the organization's resources.

PROJECT FUNDING AND RESOURCE PRIORITIES

Project size does not always directly correlate proportionally to achieving the strategic goals and objectives of the organization. So, it is possible that smaller, less costly projects may rise ahead in corporate priorities over some of the larger projects. Of course, the opposite is possible, too, as it may be more appropriate to fund the largest projects and defer some of the smaller projects if they collectively do not have the same positive impact to the organization.

OUTSOURCING CUSTOM DEVELOPMENT WORK

Many organizations outsource their IT projects and programs to external vendors. These prospective vendors can be onshore or offshore. As part of that outsourcing activity, the client may issue RFIs and RFPs). Note that, as an alternative, a request for quote (RFQ) may be sent out by a prospective customer, but in those cases, they are typically seeking to purchase a commodity item without much customization (such as when purchasing a COTS product.)

It would be a mistake to think that a vendor's response to an RFI or RFP contains accurate budget or time estimates for products or systems that are of a unique nature if they do not include within their proposals time and resources for requirements analysis and design. They should also document their assumptions that further define what they are willing to provide within the constraints of the project.

Some customers feel that vendors are overly protecting themselves when they note their assumptions in a proposal. Having been on the development side of numerous proposals, I can tell you that assumptions are necessary. Moreover, the number of assumptions escalates in direct proportion to the lack of detailed requirements information provided by the customer. In most cases, the statement of need in the form of a high-level business requirement or deliverable item is not actionable from the perspective of project scoping and project planning.

Likewise, a list of high-level requirements or deliverables from the customer does not provide sufficient detail to plan and develop a highly unique, customized solution, and support it across its life cycle. If the vendor does not believe they have enough information to understand the requirements fully, they should include budget, time, and resources for requirements and systems analysis. Moreover, the customer should expect their vendors to allocate time and resources for requirements and design analysis in their proposals.

The prospective vendors should also specify the SDLC approach they plan to take to build their customized solution. They should not hesitate to tell the customers what they need from the client to be successful, particularly when it comes to having access to customers and end users for requirements gathering, user acceptance testing, and iterative product reviews.

COTS-BASED SOLUTIONS

Many vendors provide COTS solutions that either simplify customizations or attempt to provide a common solution for all their customers. It is up to the acquiring organization to decide how much customization of the COTS product is going to be required to suit the business and end user requirements.

Moreover, if customization is required, then it should be apparent that the proposed implementation life cycle for that product or service is going to require a detailed planning effort, detailed business and end user requirements phase, and probably some development work as well before implementation can begin. As with building customized solutions, the COTS vendor should specify their SDLC approach for providing and supporting their customizations.

FIRM FIXED-PRICE CONTRACTS

Now let's talk about firm fixed price (FFP) cost contracts for a moment. When an organization issues an RFP, they may also request firm fixed pricing on the vendors' bids. FFP means the client is asking the vendor to guarantee the cost of designing, building, and delivering the solution. Moreover, the client is asking the vendor to assume all the risks for delivering the entire scope of work detailed in the RFP.

Going back to our previous discussion—that a list of deliverables does not provide sufficient requirements to work from—it should be apparent that the vendor assumes a substantial risk if they do not include time and resources for requirements and design analysis. Truthfully, there is also a risk to the client since the vendors do not have a complete understanding of the client's needs and requirements, and as a result may propose the wrong solutions.

Conceptually, an RFP is different from an RFQ in that the RFP describes a need for customized products that have a high degree of uncertainty in requirements, pricing, and time. In contrast, an RFQ specifies a client's desire for relatively standard products, commodities, or COTS offering.

Project management is the best approach to managing the development of highly customized or unique products. Therefore, an RFP request and the vendors' proposal responses should include a life cycle development process that encompasses the scope of work identified within the proposal. That is true whether the development team uses traditional or Agile-based SDLC practices.

DID YOU KNOW?

Under an FFP contract, the risk of delivering the scope of work within time, quality, and budget constraints falls to the vendor.

However, even under an FFP contract, the client is not without risk.

It is the client's responsibility to identify and provide the requirements or provide sufficient access to those who understand the requirements, and to ensure that the project properly addresses the strategic goals and objectives outlined in the client's mission statement

If a vendor's proposal does not include a proposed SDLC, that should be of concern to the client. Without a statement of the planned SDLC, the client cannot be sure the team plans to commit sufficient time and resources to plan the project, gather and analyze requirements, and design the right solution and architecture. Moreover, the client does not have the information they need to trust the vendor's estimated delivery dates.

Some clients may view vendor proposals that have added time for project planning and requirements and design analysis as being overly expensive, or as coming from a vendor who doesn't understand what they're doing. But those vendors are reducing the risks for both the vendors and the client alike. Assuming the proposals are coming from qualified vendors, those are the proposals that should be taken most seriously.

You may wonder why a vendor would outline their development strategy on an FFP proposal. After all, they are only on the hook for delivering the final product. However, the customer may be lulled into believing the vendors have everything they need, information-wise, to fully assess the scope of work, budgets, time, and resources required to produce the deliverables. The FFP proposal that outlines the vendor's SDLC approach is also outlining the vendor's requirements from the customer to ensure the overall success of the project.

Let me say this as succinctly as I can: requests for proposals almost never provide a complete set of requirements. Therefore, even under an FFP contract, the client must understand that they need to provide the client access to their domain and SMEs to support the vendor's requirements gathering and analysis activities. The client must also support reviews of prototypes and iterative releases.

Vendors must protect themselves when submitting a proposal by including within their bids the time, resources, and budgets to complete full requirements and systems analysis. The vendor should also request customer reviews at the end of the requirements and design phase, after completion of any prototypes, and after each iterative release of the products under development.

Dangers in Shortchanging the Life Cycle Process

Sometimes vendors shortchange the development life cycle process to be cost competitive and win the business. This is not a good practice for either the vendor or the client, as eventually the vendor will be exposed.

The vendor who does this may be betting that the client will award additional funding as the requirements are fully defined and understood, but they are taking a risk. The client might instead simply choose to cut their losses and pull the contract, or they may force the vendor to deliver the entire scope of work to the time and price constraints agreed upon in the original contract. In fact, under an FFP contract, the client will insist on completion of the entire scope of work. They have a legal right to do so.

Vendors who assume they have the right solution for the client but are unwilling to bid proper time and resources for project planning and requirements and systems analysis are going to be the vendors who are most subject to the pressures of scope creep. In this case, scope creep can occur when the customer keeps asking for features and functions that the vendor does not believe were included in the original contracts or project charter.

Scope creep occurs in internal development projects for these very same reasons. The work performed in a business case, ROI, or portfolio analysis is not a substitute for the detailed project scoping work conducted under the project planning and requirements and design analysis processes. It's better that the executive sponsors break out the initial project scoping work as a separate project, and then use the resulting information to derive and approve the constraints for the product development project.

CHAPTER SUMMARY

In this chapter, you have learned that the executives of an organization use strategic planning to align the business with its vision and mission. The strategic plans often lead to the implementation of change initiatives within the enterprise. There are many factors that can drive organizational change, and such change must be managed. We also learned that standard life cycle processes help organizations implement and manage change.

07.
Corporate IT Standards

This chapter will be useful for those readers who want to understand why corporations should implement corporate IT standards, document best practices, and continuously improve their SDLC processes..

CHAPTER OBJECTIVES

After reading this chapter, you should be able to discuss the following:

- Explain the role of IT governance boards and frameworks.
- List two types of IT standards.
- Explain the value of employee SDLC best practices and a maturity process.
- Explain how proven and repeatable methods and tools improve life cycle development project outcomes.
- Explain why documentation is such a contentious issue with some Agile practitioners, but also articulate why some level of documentation and record keeping is needed.
- Describe how legislative requirements drive IT governance policies and processes for financial and healthcare-related record keeping

AUTHOR'S INTRODUCTION

So far, we have discussed foundational software and SDLC approaches and common SDLC methodologies, the impact of organizational change, and the need to perform some level of scoping work to better inform the change requirements identified in strategic planning. I've also discussed the use of systems and PMLC processes to help the organization manage their strategic change initiatives.

Along the way, I've discussed the need for the organization to establish IT standards. Standards come and go, and not all IT standards are relevant to every organization. This chapter discusses the need for the establishment of IT governance policies and IT standards, documentation of best practices, and the need to seek continuous improvements through process maturity programs. These strategies help the IT organization achieve consistency of outcomes while reducing risks and the number of failures.

One thing to note about standards is that they can accumulate like old books on a bookshelf. They get old, dusty, and musty. They need to be dusted off and reviewed if you are to receive continued value. Moreover, just like a book that requires updating over time through new editions, standards likewise are updated or replaced. Plus, as new technologies and methods evolve, new standards emerge.

The IT policies and standards approved for use within the organization must be maintained and updated over time. The oversight of this work falls to a group of executives who make up an IT governance board, and they look across the needs of the organization, independent of the IT organization, to ensure that IT investments are appropriate, supportable, and sustainable. In addition, IT governance plays an important role in implementing IT controls and reporting to support compliance with legislative requirements.

The implementation of IT governance policies and IT standards can contribute to documentation requirements throughout an IT project and accumulate over the life of the product. The amount of SDLC related documentation is a point of contention among advocates of Agile practices. Not that Agile practitioners don't believe documentation is required; rather, they question how much documentation is required.

This chapter discusses the reasons some level of documentation is always necessary. The documentation burden can be reduced by using documentation templates and various life cycle systems development tools that support capture, management, and reporting on product life cycle information.

Finally, the chapter dives further into the need to develop mature SDLC processes that are consistent with the IT governance policies and IT standards approved for use within the organization. In effect, the SDLC processes transform the IT governance policies and standards into an actionable set of processes that can be applied repeatedly and consistently across the enterprise.

SDLC STANDARDS

In the course of this chapter, you will be exposed to countless standards governing various aspects of information technology. But, in this introductory section, let's start with the most basic.

Over time, efforts have been taken to codify current SDLC approaches into international standards. The latest is the joint ISO/IEC/IEEE 12207:2017 Systems and software engineering -- Software life cycle processes, which is an international standard for software life cycle processes.[84] Another current international standard is ISO/IEC/IEEE 15288:2015 Systems and software engineering -- System life cycle processes. [85]

ISO/IEC/IEEE 15288:2015 has a focus of providing process descriptions for systems-oriented development life cycles and doing so from an engineering point of view. In contrast, ISO/IEC/IEEE 12207:2017 provides a framework encompassing full software life cycle processes with related terminology. Both standards provide a defined set of processes and terminology to facilitate communication across the international IT community, including those who are acquiring and supplying software and systems, plus other interested stakeholders.

IT GOVERNANCE BOARDS

One of the roles of IT governance is to determine the IT standards that best apply to the needs of the organization. An IT governance board typically has executive level appointees from across the business enterprise. Their role is to provide unbiased assessments and decisions on IT policies and investments with the primary focus on supporting business objectives. The cross-organizational structure and executive level membership prevents an IT organization from overly influencing IT investment decisions that are not in the best interests of the other stakeholders across the enterprise.

That's not to say that the IT organization does not have input on IT governance decisions. It would be foolish to exclude the IT experts from decision making on policies, standards, and technologies. The IT governance board should solicit input from the IT organization's expert and may have technology-focused sub-boards that report to the IT governance board on matters of key interest.

In my experience with IT governance boards, the executives typically meet monthly to decide key issues and provide direction to the technology sub-boards. The technology sub-boards meet more frequently, even weekly, to work on IT policy, product, enterprise architecture, and technology assessments, and review and update as necessary the organization's IT processes.

IT GOVERNANCE FRAMEWORKS

As with the SDLC framework, IT governance is considered another type of critical support framework. Where an SDLC framework provides structure to manage the organization's IT processes, the IT governance framework provides structure to ensure that the organization's IT policies, standards, and investments support its business objectives.

A critical objective of IT governance is to ensure that the organization is meeting regulatory compliance in both the capabilities of their IT systems and the appropriate application of their IT systems to meet regulatory requirements in other areas of the business.

An organization can develop an IT governance framework from scratch, but it's typically more effective to draw from an existing standard IT governance framework. Some examples include:

- AS8015-2005—Australian Standard for Corporate Governance of Information and Communication Technology
- CMMI
- CobiT 5—originally Control Objectives for Information and Related Technologies
- COSA Framework—Committee of Sponsoring Organizations
- FAIR—Factor Analysis of Information Risk
- ISO/IEC 38500:2008 Corporate governance of information technology
- ITIL 4 IT Service Management (ITSM) framework (originally Information Technology Information Library)
- Val IT 2.0—Value Management Guidance for Assurance Professionals

In the broadest sense, corporate governance deals with how a company is managed and directed, most directly in the interests of the shareholders, but also in the interests of other stakeholders such as employees, investors, creditors, suppliers, customers, and affected communities. The role of IT governance boards is to ensure that senior executives are establishing policies and procedures and making critical decisions on technology, support, and investment decisions.

Governance programs also implement controls within a corporate environment to ensure compliance with legal and legislative requirements.

IT governance policies set the path by which the organization remains compliant with legal requirements and legislative mandates. The IT governance policies help keep the organization and its executives out of trouble. The SDLC framework maintains the SDLC processes, activities, tools, and templates that support the organization's governance policies.

Without getting into too much detail, the legal requirements for corporate governance evolved after several high-profile corporate failures resulted due to fraud and mismanagement by senior executives. In the United States, the Sarbanes-Oxley Act (SOX) has been the principle legislative requirement implemented to ensure that proper controls, monitoring, and transparency exist for financial reporting within a public company.

Sarbanes-Oxley (SOX) Act of 2002

This legislative act, also known as the Public Company Accounting Reform and Investor Protection Act (in the Senate) and Corporate and Auditing Accountability and Responsibility Act (in the House), establishes a set of 11 sections that detail responsibilities for corporate executives and board members, plus criminal penalties for misconduct. The primary emphases of SOX are on financial controls, transparency, and accountability.

SOX does not directly address information technology controls; however, Section 404 requires chief executives and financial officers to confirm the accuracy of the firm's financial reports, and the company must provide internal controls over financial reporting. [86] Failure of the company's executives to ensure adequate controls over corporate data can result in severe legal fines or even criminal charges.

Since much of a public company's financial records are maintained on computing systems, Section 404 requirements apply to the internal controls implemented via the corporation's financial systems and related testing to ensure that the company and its officers have not accidentally or intentionally misstated the financial records of the company.

CobiT is an example of an IT governance framework that helps the organization support legislative requirements, and specifically those mandated under SOX. COBIT provides an internal control framework used to measure and audit proper IT governance and control to achieve SOX compliance in IT.

Healthcare Related IT Legislation

The healthcare industry is another business area that is highly regulated, with a resulting impact on the implementation of information technology policies, processes, and systems. The healthit.gov website provides information on Health IT legislation.[87] The current list of Health IT legislation includes:

- 21st Century Cures Act—Provisions to improve the flow and exchange of electronic health information.
- MACRA—The Medicare Access and CHIP Reauthorization Act of 2015 is a two-track system that requires physicians and clinicians to accept downside risk while offering payments under a merit-based incentive payment system (MIPS) or the alternative payment models (APMs) under Part B (medical insurance) of the Medicare program.
- HITECH Act—The Health Information Technology for Economic and Clinical Health (HITECH) Act of 2009 establishes programs to improve health care quality, safety, and efficiency through the promotion of health IT.
- FDASIA—Section 618 of the Food and Drug Administration Safety and Innovation Act proposed a strategy and recommendations for a risk-based regulatory framework for health IT, including mobile medical applications, that promotes innovation, protects patient safety, and avoids regulatory

- **HIPAA—The Health Insurance Portability and Accountability Act (HIPAA) Privacy Rule describes what information is protected and how protected information can be used and disclosed. The HIPAA Security Rule describes what safeguards must be in place to ensure appropriate protection of electronic protected health information.**

- **Affordable Care Act—Promotes the use of health information technology (HIT) to study outcomes of treatment protocols, assess risk factors associated with disease processes, assess the overall health of specific populations, track effects of certain medications over time, and collect and analyze health-related data in other areas of concern.**

Regulatory Impacts on IT

I've included the previous two sections on SOX and HIT legislation to make two points. First, legislation impacts the business of business, and IT is an important enabler of meeting compliance requirements. Where SOX legislation is intended to protect shareholder and stakeholder interests on financial matters, HIT legislation seeks to similarly protect and regulate consumers, patients, healthcare providers, and medical suppliers.

Second, SOX and HIT legislation is only a small subset of the legislation that can affect business. An organization's legal representatives need to work with the IT governance board to ensure that they have a thorough understanding of legislation that can affect the business and implement IT policies and systems that ensure organizational compliance with relevant regulations.

One of the interesting dichotomies of SOX is that while it provides significant financial and even criminal penalties for failures to implement internal controls properly and to ensure accurate financial reporting, it never specifically mentions information technology as the means to comply. Likewise, SOX does not identify IT standards that might be adopted to demonstrate compliance. This is true of virtually all legislation. Legislation tends to direct data collection and reporting requirements and data sharing restrictions but does not dictate how organizations are to implement the mandates.

In this light, public companies must choose the IT framework they believe will best ensure compliance with legislative requirements that affect their business. For example, within the financial industry, audit firms often choose the control objectives for information and related technology (COBIT) framework as the preferred standard for their industry.[97]

Other frameworks can be considered, such as information technology infrastructure library (ITIL), CMMI, managed object format (MOF), and the open group architecture framework (TOGAF). The main point to be made here is that IT governance frameworks are useful for managing IT projects in a manner that support the needs of the business while also helping ensure regulatory compliance.

IT STANDARDS

In my research for this chapter, I decided to do a quick search on IT standards in existence today. A search on Google brought up lists of 68 hardware-oriented standards, 59 software-oriented standards, and 111 networking standards.[98, 99] Not that I'm implying Wikipedia is a definitive source for this type of information. This look up was more to gauge how many standards potentially exist, as documented by those folks who care to track such things.

Hardware standards govern the use of computing equipment, such as servers, laptops, desktops, printers, and mobile devices. At an organizational level, software standards can apply to the use of specific custom applications or COTS-based software authorized for use by an organization. However, in a broader sense, software standards cover the technology standards related to developing software systems. Networking standards define the rules of communication between various networked devices.

Hardware, software and networking standards are not the only categories of IT standards.

Other examples include standards for storage, middleware, geographic information systems (GIS), security, web development, and others.

Another source of IT related standard is IT Governance USA Inc.[100] They offer standards published by the International Organization for Standardization/International Electrotechnical Commission (ISO/IEC), the International Organization for Standardization (ISO) and the British Standards Institution (BSI). They list 60 ISO/IEC standards for information security and IT service management in eleven categories.

You may wonder why there are so many additional IT standards compared to the number of ISO/IEC standards. As it turns out, there's not much overlap between the two lists of IT standards. The ISO/IEC-based IT standards support IT governance and introduce related terms, definitions, and processes. In contrast, the Wikipedia article lists IT standards that are technology oriented.

Where the ISO/IEC IT standards are process-centric approaches to employ IT in support of organizational business objectives, the other IT standards are related to engineering practices. Technology-oriented standards provide guidance on technical or engineering processes, protocols, and practices. Formalized standards are developed through an accredited agency following a consensus-based decision-making process.

BEST PRACTICES AND MATURITY PROGRAMS

Without a doubt, the most important outcome of implementing a viable corporate level SDLC is to encourage the establishment and use of the organization's best practices. Those best practices can be engineering-oriented or process-oriented.

It's not enough to hire smart people. By analogy, a football team can have a lot of talented players and yet fail to win any games. Coaches develop a set of their best practices, documented in the form of a playbook, train their players in their use, and expect their players to execute the plays faithfully when called. One can think of an SDLC framework as the playbook for the IT organization.

The development of a corporate SDLC standard should be part of the organization's evolving maturity/process improvement program. A mature SDLC framework defines life cycle processes, methods, tools, activities, tasks, milestones, and metrics in the form of acceptance criteria. Developing this information from scratch is not a trivial endeavor. It takes time and resources.

On the positive side, much useful information already exists in the heads of the experts within the company and is informally shared among team members. Improved consistency in project outcomes, reduced risks, and improved efficiencies comes from capturing the best practices of the experts and making it available to all project teams. The trick is to make the capture of that information as painless as possible. The Activity Worksheet presented in my books titled *Frameworks for Agility* and *Tools and Templates* aid in the capture and deployment of best practice information.

Consistency of Outcomes

The previous subsections spoke to the establishment of best practices and the use of proven and repeatable methods. The entire reason for doing so is to improve the consistency of outcomes for each development project while ensuring a higher overall level of success. Improvements in quality, cost containment, customer satisfaction, and revenues from the development of new products and services, when and as needed, contribute directly to higher project success rates.

These objectives might sound more like operations-oriented goals as opposed to project-based goals. It is true that in the operations side of the business, numerous methods have evolved to establish and implement repeatable processes, with the goal to achieve predictable outcomes controlled within narrow bands of performance metrics. Examples of operational efficiency methods include JIT production, Kanban-based scheduling, six sigma for process improvements and defect elimination, and lean manufacturing.

In an operations-oriented environment, the primary objectives are to control costs and wastes. The organization must be able to do basic things repeatedly with great accuracy and predictability and be able to produce products in step with demand.

These techniques are finding a home in IT, even in the execution and management of project-oriented work. Also, just because a project is developing a unique product within a specific time limit doesn't mean there can't be reuse of best practices, methods, and tools across each product life cycle. Life cycle processes and best practices help the organization control risks through reuse of proven methods.

SDLC DOCUMENTATION REQUIREMENTS

Most advocates of Agile software development methodologies abhor excessive amounts of documentation; nevertheless, some level of documentation on every system or product development project is required. However, before we get too far into this conversation, it's important to address the concern that Agile practitioners have—that excessive documentation takes time and resources away from developing a product. In their view, less gets done and at more cost.

Having said this, executive sponsors and customers have other critical concerns that must be balanced against the time and effort it takes to provide detailed documentation. Those who are paying for the project require commitments from the development team on what capabilities will be implemented, as well as assurances that the work can be accomplished within the allotted constraints of the project. In other words, the development team must perform some level of analysis and documentation to ensure that the client gets what they expect for the time and resource made available to the development team.

Once the solution is developed, it must be sustained, supported, and possibly extended. This means that the maintenance staff requires systems and data administration, installation, and recovery documentation. Future developers need requirements, architecture, and design documentation so that they know how and why the system was built, and how to extend or modify the system. The support staff needs information on how the system works in order to have the information they need to answer end user questions. Affected business processes require new documentation so that the employees and management know how to do their jobs, and end users need training on the new system, which requires the development of Wikis, user guides, and other training aids.

My *Building our Digital World* Series Book titled *Tools and Templates* introduces more than 100 templates that can be applied across the life cycle of a software product or system. No project should expect to use but a small fraction of the available templates. Some are not relevant. Other templates may not have sufficient value for a project to justify the time and effort to use them. The resources and time needed to complete a template must always be balanced against the benefits.

So, while the organization may maintain all the templates for deployment and reuse within an SDLC templates repository, each project team must meet with their sponsors and agree on the amount and type of documentation that supports their client's information requirements. Some templates may be mandated across all projects.

Finally, some modern development tools and suites provide data capture and reporting, collaboration, and information sharing capabilities throughout the product development life cycle, greatly reducing the documentation burden. A modern and full life cycle IDE should have these capabilities.

CHAPTER SUMMARY

In this chapter, we have discussed the need for an organization to establish IT governance boards and frameworks. The IT governance boards develop policies and standards that support stakeholder and business needs IT governance policies and standards to help ensure that the organization implements technology, support and investment decisions consistent with the need of the business and not overly influenced by the IT organization and practitioners.

IT governance also helps the organization achieve compliance with legislative mandates. For example, SOX and federal health IT legislation were introduced as representative examples of government legislation that mandates specific requirements in the area of financial matters and healthcare.

08.
Expanding Our View of Systems

This chapter explains how why the contextual use of the SDLC acronym has evolved to include software, systems, and solutions, and how life cycle development concepts can apply to help manage change across all enablers of the business enterprise.

CHAPTER OBJECTIVES

At the end of this chapter, the reader should be able to discuss the following:

- What are the three related definitions for the acronym SDLC?
- What is the purpose of an SDLC framework?
- What are the common phases or stages of work?
- How do Agile-based SDLC processes fit within an SDLC framework?
- What is the difference between SDLC frameworks that allow multiple process paths versus those that employ hybrid life cycle models?
- What is the purpose of hierarchically structured SDLC frameworks?
- How are the process and task sequencing implemented within the SDLC, project planning, and when developing project schedules?
- How do SDLC frameworks facilitate reuse of standard methods and tools?
- What is the value of using standard SSDLC forms and templates?
- What are business-oriented development life cycles and how do they impact the evolution of IT systems?

AUTHOR'S INTRODUCTION

I noted in Chapter 1 that SDLC originally stood for *software development life cycle* but now has come to more commonly mean *systems development life cycle*. The primary focus of this chapter is to expand our thinking on what a "system" might be.

One of the reasons the term "system" has replaced the term "software" is that software, and IT in general, have become embedded components in almost any product we can imagine today. The collection of computing-enabled and highly integrated products certainly embodies the concept of more complex systems. In fact, I would argue that the concept of the Internet of Things (IoT), where computing devices are embedded in common items and connected to the Internet, has expanded the concept of a system to include any number of interconnected devices over the Internet.

Businesses operate as highly complex systems, with or without IT. However, since this is a book on SDLCs, businesses also operate as a form of an integrated system when management employs IT to integrate, automate, and support critical processes.

In this book, I've presented a case that project management and development life cycles are both instruments for managing change. PM processes facilitate the management of unique and temporary work endeavors. Standard development processes provide a standard approach to creating items that are similar but sufficiently unique to make the work difficult to operationalize.

I'm not saying it's impossible to operationalize development of unique or customized work products. In my book titled *DevOps*, I discuss lean, JIT, Kanban, and continuous integration, testing, delivery, and deployment practices. These techniques help operationalize work even when the work products are unique from one instance to the next. For now, it's important to understand that mature PMLC and SDLC processes work in concert as aids to help the organization create novel IT products and systems.

In this chapter, we are going to find out that project management and life cycle processes also help manage change within and across the enterprise. I also present a conceptual framework upon which various project management and SDLCs can hang. Since this is not a book on management consulting, this chapter will bring the discussions back to the impact of organizational change on IT.

SDLC TERMS AND DEFINITIONS

So far, I have used the terms software, systems and solution interchangeably, and for the most part, I'll continue to do so. However, in this chapter, I need to define these terms explicitly, especially as they apply when using the SDLC acronym.

To start, while software development is a risky endeavor for most organizations, pure software development projects represent only a portion of an entity's change investments. Many other initiatives involve information technology as part of a larger, complex, and integrated system. For this reason, the phrase "software development life cycle" is too limiting a term for most IT-based development projects.

The word "systems" replaced the word "software" in the SDLC acronym over time as it became apparent that software is a small component of many IT projects. Moreover, infrastructure investments in computing equipment, networks, security, and storage components could be every bit as complex and expensive, if not more so, than the expense of developing software.

Moving forward in time, software and related computing and communications technologies are finding their way into almost every product we use today. So, a system by definition can include things like televisions, smartphones, automobiles, aircraft, home security systems, appliances, robotics, you name it.

Some organizations have used the word "solutions" to replace the words "software" or "systems" in the SDLC acronym. The word "solutions" brings to mind the notion that products and services have limited value if they are conceived and developed without the customer in mind. In other words, customers need solutions to their problems.

SDLC FRAMEWORKS ORGANIZE WORK

Regardless of whether the SDLC acronym refers to software, systems, or solutions, applicable to all is a common conceptual model for life cycle processes. For example, the traditional systems development model typically specifies distinct phases or stages of related work. This practice followed the traditional linear-sequential waterfall approach to development.

Though different versions of such SDLC models may have different phase names or a different number of phases, they usually have more in common than not. For this book, I have adopted phases, names, and descriptions that are in alignment with an approach used by SDLC Forms, a company I've collaborated with over the past three years as this book was evolving. The phases of our conceptual SDLC model include:

- **Concept**—In this phase, the organization documents best practices, methods, and tools to determine the feasibility of a proposed initiative, perform ROI analysis, and build the business case to justify the project. This phase of work may also include a CONOPS investigation, which is a mission-oriented capability needs assessment from the perspective of end users. In other words, a CONOPS study investigates what capabilities a system must provide to assist end users in their work that supports the mission of the enterprise.

- **Requirements**—In this phase, the SDLC provides best practices, methods, and tools to conduct business needs analysis, business process analysis, and user needs/usability analysis. The deliverables with this phase of work are a business requirements document, or use cases, or epics and user stories. The requirements identified in this phase are more refined than those found in a CONOPS study.

- **Design**—Typically includes systems analysis, prototyping, architecture, and design activities. Systems analysis looks at requirements from the perspective of what the system needs to be able to do to meet business, process and user requirements. Prototypes serve as a proof of concept and provide an example to solicit customer and end user feedback. The system architecture is a high-level description of the primary components of a system, and the technology standards, methods, and tools used during development. Design activities get into the details of constructing the individual software components.

- **Development**—In this phase, the SDLC provides technology, engineering, and best practices to develop the software and system components. Development is highly dependent upon the IDE, or software development tools used by the organization. However, for more complex systems, this phase includes the descriptions of the methods and tools in the construction or implementation of ancillary solution components, such as hardware, network, reliability, backup, and security systems.

- **Testing**—The methods and tools used in this phase of work, as with the development phase, are highly dependent upon the technologies chosen for development and testing. Concurrent with the software development activities, the development team conducts unit testing to ensure that each section of code they create meets the criteria specified in the business, process, user, and design requirements. Integration testing is also performed concurrently with development to ensure that each new component of code works with the applications existing code. After development, the software solution components are integrated and tested as a whole through a series of quality assurance, user, performance, regression, production, and potentially other testing procedures.

- **Production Turnover**—This is the phase where the product is turned over to the customers and user communities. However, the work starts with implementation and transition planning plus deployment related activities. In other words, production turnover activities occur in parallel with the design and development of phase-based activities. The production turnover process includes work to create user and systems guides, application procedures and policies, transition plans, job aids, wikis, or any number of other artifacts to help ensure that the system is operated and maintained successfully after deployment release of the systems on the production (e.g., target) environments.

- **Postproduction/Operations & Maintenance (O&M)**— This is the phase where ongoing product and user support is required. Also, the system may continue to evolve, requiring further enhancements, testing and bug fixes. When the product is eventually replaced, end-of-life activities may be required to dispose of system components that present an environmental hazard. Also, end-of-life activities typically involve the removal and archiving of application data, and the proper destruction and disposal of computing equipment and storage components.

Other developers of traditional SDLC frameworks with the phase-based categorization of work may use different names or even a different number of proposed stages. However, there is more similarity than differences between the names and descriptions of the categories of work.

PHASES OR STAGES OF WORK

The SDLC phases depicted in Figure 16 follow a traditional life cycle development approach. As shown in the figure, the top-down structure of the phases might suggest that developers start at the top and work our way down through a series of distinct phases, in more a less a sequential manner. However, in modern practice, the phases overlap and are often repeated, particularly when using Agile practices.

In the traditional hierarchical SDLC model, the SDLC framework provides the structure to manage information based on types of work. An IID-based SDLC combines requirements, design, prototype and development activities into one or more distinct iterative development cycles. Each iterative development cycle repeats multiple times per release but under shorter time durations than is typical under waterfall approaches.

Systems Development Life Cycle (SDLC)

Life Cycle Phases	Life Cycle Phase Descriptions
Concept	Determine feasibility, cost benefits, concept of operations, project scope
Requirements	Gather, analyse and dicument business & functional requirements as use cases of epics, these and stories
Design	Includes prototyping, systems requirements, architecture and design
Develop	Building the solution's components; with software, may incorporate Agile methods
Test	Unit, component, systens, integraion, regression, quality assurance, user acceptance, performance, production, and other types of testing
Production	Focus on readiness for deployment, including development of various guides/aids and training
Deployment	Prodcut and user support, enhancement requests, bug fixes, disaster recovery, end-of-life

Figure 16. SDLC Life Cycle Phases, A Conceptual Model

Nevertheless, an Agile-based SDLC model does not negate the importance of having a phase-based structuring of SDLC processes, activities, and tasks. In effect, the Agile model overlays the traditional model repeating the processes iteratively and more frequently than under the traditional SDLC model.

Agile SDLC Processes within an SDLC Framework

The organization and graphical views of Agile process models often look different than the traditional linear-sequential models. For example, Agile models that implement IID practices have graphical models that depict spirals or loops in construction. Agile models that employ lean and JIT practices employ graphical views that show the limitations placed on work in progress.

There are other fundamental differences between the traditional and Agile SDLC models, particularly in their approach to describing work. For example, although the traditional waterfall model implied the linear-sequential deployment of life cycle development processes, the real value is the establishment of a framework to categorize and manage work. In contrast, the graphical views of Agile are designed to show the flow of work in the iterative or limited WIP SDLC models.

In the traditional SDLC model, development processes are described once, and then managed and accessed under the appropriate category of work. The structure of the traditional SDLC model is hierarchical (see Figure 17), conceptually similar to the drawers, folders, and files found in a physical filing cabinet or the directory tree and files structure of an electronic filing system. As a result, the traditional model provides a useful structure for organizing and managing SDLC process information.

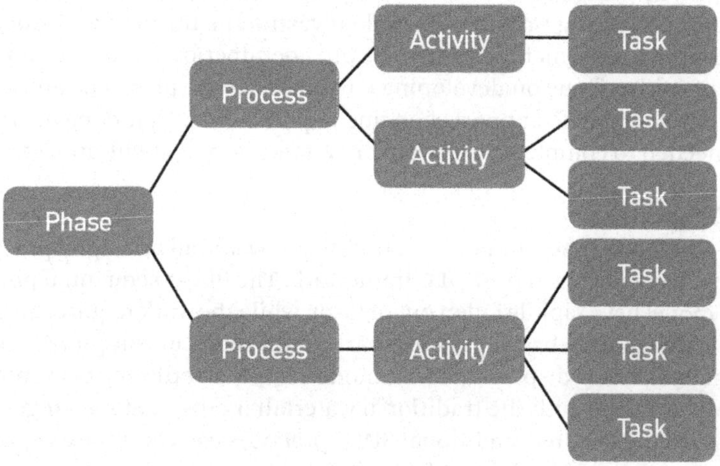

Figure 17. Hierarchical Structure of the Traditional Phase-based SDLC Model

The graphical structures in Agile models do not provide an efficient mechanism to manage process information at a granular level. The focus of Agile models is on describing efficient processes, and not on approaches to manage, retain, and deploy SDLC process information and best engineering practices.

Multiple or Hybrid Life Cycle Paths

I've been fortunate to have had the opportunity to work with several corporate standard SDLCs that have incorporated multiple SDLC process paths or hybrid SDLC approaches. With some upfront work to make proper assessments, a development team can make either approach work.

An SDLC framework with multiple paths provides alternative sequences of processes that the project team can select from as they build their project plan. A hybrid SDLC model incorporates both traditional and Agile practices as separate paths and allows the team to choose which of the two models best support their project needs. The DA Toolkit is a blend of the two concepts, with any number of practices included in the SDLC framework and multiple implementation paths. The transition from the traditional waterfall model to Agile has been difficult for many organizations. Some of the early SDLC models implemented by my customers were ill-conceived and didn't work out in the long run—especially those that forced iterative waterfall approaches. Others provided more flexibility in approaches and supported tailoring and selection of SDLC processes and Agile approaches to meet unique project-level needs. Based on my experience, those types of hybrid-SDLC approaches seem to work best.

For example, the traditional waterfall model, with its structured requirements gathering, analysis, design, and documentation practices, tends to be useful when the business requirements are difficult to break down into relatively small chunks that fit within a development iteration. Also, sometimes the requirements are so large and complex that the work is difficult to analyze properly before starting the iterative development cycles. Other projects, particularly smaller projects with simpler requirements, can go straight into an Agile framework incorporating prototyping and IID practices. Plus, lean concepts are proving useful to eliminate wasted efforts and the clutter of work that can derail a project.

Structured SDLC Frameworks

Based on my experience, a hierarchical SDLC framework simplifies management of SDLC process information and engineering practices. The goal should be to develop a hybrid SDLC framework that doesn't force the inefficiencies of the traditional model upon Agile development practices. Rather, the SDLC framework provides the structure for organizing information, and pulls only those processes and practices into a project plan that support the needs of customers and the project team.

It may seem odd that traditional and Agile SDLC processes can be blended. The traditional waterfall development practices tend to have a greater focus on initial requirements and design analysis and protecting product life cycle investments through CM, documentation, and training. In contrast, Agile methods and tools provide superior approaches for efficient and rapid development, with a focus on developing customer-centric products, and offering more flexibility during development to support changing requirements. To varying degrees, both sets of processes are needed to complete a software or systems development project.

Sequencing SDLC Processes

The execution of SDLC processes, under both traditional waterfall and Agile practices, should not be enforced by the structure of the SDLC framework. The linear-sequential processes of the traditional SDLC model have rapidly fallen out of favor, while the work required in each iterative development cycle under Agile-based practices varies with then-current project priorities.

At a conceptual level, visual displays of SDLC models imply an ordering of events. (See Figure 18) However, as we have seen with the traditional waterfall model, that's a dangerous practice. I've already discussed the fact that traditional SDLC practices can combine a rigorous phase of initial requirements and design analyses before initiating prototyping and iterative development and testing phases.

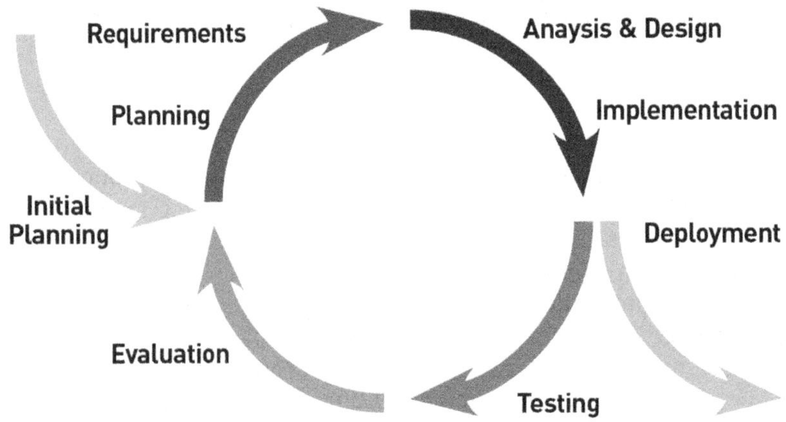

Figure 18. Iterative and Incremental Development Model

Likewise, the visual models showing repeating iterations under various Agile-based SDLC models are overly simplistic. In large and complex systems development projects, the type of work conducted in each sprint or release can vary significantly. In other words, while work can be structured into smaller and repetitive iterations of work, the actual type of work and SDLC processes required can vary significantly in each iteration of work.

Therefore, regardless of the SDLC framework employed by an organization; it is not the order of placement within an SDLC framework that determines the flow of work. Rather, it is the inputs and outputs of individual processes, activities, and tasks that drive the sequencing of events through an SDLC. Project managers have known this for a long time. In the next section, I'll begin to explain how project managers approach project planning and scheduling. Then, we're going to take a much deeper dive into this topic in Chapters 11 and 12.

PROJECT PLANNING AND SCHEDULING OF WORK

The focus on project planning, whether under traditional or Agile development practices, is to define the total scope of work associated with creating the deliverables approved for the project. The processes specified within a corporate SDLC, with their inclusion of information on standard methods, tools, resources, skills and estimated duration, provide a sanity check on the team's ability to build a product within allotted budget, time, resource and quality constraints. The project plan and project schedule also provide the information necessary to track whether the project is on track to deliver the final deliverables on time and budget.

Whether using a traditional development life cycle or Agile, the inputs and outputs of the SDLC processes provide the sequencing instructions necessary to create the project schedule. Many of the underlying methods remain the same across each development and test iteration. Certainly, that is true for the organization's preferred technology and engineering-based practices and tools.

It is much easier to schedule specific work and associated deliverables out into the future when the project team includes a detailed requirements analysis and design phase before implementing iterative development practices. However, under a classic Agile-based SDLC methodology, the deliverables are not defined at a granular level in the early stages of the project. In those cases, the project manager schedules future sprints and releases in advanced as placeholders. As the project evolves, the project managers fill in the blanks with the establishment of the product backlog of desired features and functionality, based on development priorities.

The hierarchical SDLC model, listing reusable SDLC processes under types of work simplifies classification and management of the organization's approved processes. When creating a project plan, the project team can easily find and pull the information required into their project plans and schedules. The dependencies of each project task form the schedule.

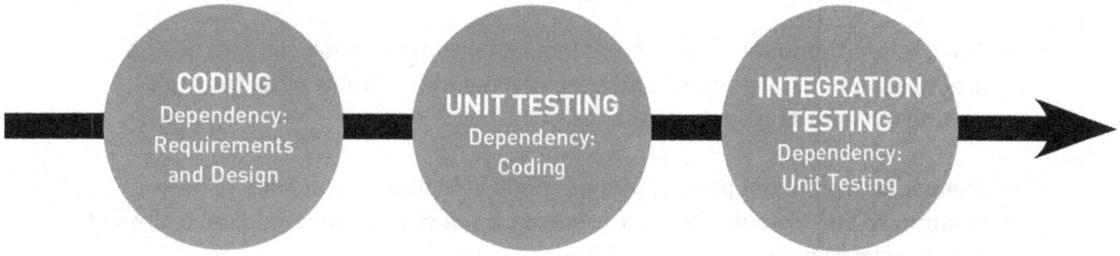

Figure 19. Coding, Unit Testing and Integration Process Flow

So, for example, unit testing logically follows coding, and integration testing logically follows unit testing. Those are easy dependency relationships. (See Figure 19 – Coding, Unit Testing and Integration Process Flow for graphical depiction.) However, when is it appropriate to perform regression testing, smoke testing, or performance testing? The decision criteria and dependency relationships for those testing requirements are more difficult to represent in a visual SDLC model. However, the parameters for those decisions can be easily documented in the organization's approved SDLC processes as dependencies, thereby making it easier to recall and incorporate them in the overall project schedule.

REUSE OF METHODS AND TOOLS

For the most part, the technical tools developers use for constructing a product or system remain the same in both the traditional and Agile SDLC models. For example, coding methods remain the same under both traditional and Agile development approaches and are primarily dependent upon the technology authorized to code the software. Unit testing is still unit testing, regardless of the SDLC approach. Likewise, the methods and tools used for integration testing do not need to change, except that integration testing is performed more frequently under Agile development approaches.

Under Agile, requirements are gathered and analyzed over time, with the design, development, testing, and customer review processes combined and executed across multiple short-term development iterations. An Agile-based project schedule must likewise show multiple planned development iterations.

Given the evolutionary development process of Agile, the schedule has more ambiguity regarding the specific deliverables associated with each development iteration. Nevertheless, the implementation of Agile practices does not have much effect on the underlying methods used for designing, developing, testing, and deploying a product. They all just happen more rapidly and frequently.

To recap; processes, methods, and tools from both traditional and Agile-based approaches can be blended to allow:

- The use of traditional SDLC processes when the product requirements and designs are too large and complex to easily resolve through iterative development practices.
- The use of traditional SDLC processes to help protect corporate IT investments through documentation, training, and CM practices.
- The employment of Agile SDLC processes to support clients with changing requirements or when the understanding of requirements needs to evolve with the product.
- Implementation of prototyping and iterative development practices to allow more frequent product releases driven by customer feedback.
- The use of traditional project planning, requirements analysis, and design practices to help customers and executive sponsors know what they are getting for their IT investments.
- The use of traditional project planning and scheduling practices to help customers and executives track progress against the approved constraints of scope, schedule, budget, and quality.

STANDARDIZED SDLC PHASES, FORMS, AND TEMPLATES

There is not a single approach to describe a standard corporate SDLC, nor will other corporate SDLC approaches use the same names I have used in this book. The phases described in this book are fairly common in the IT industry. Some organizations may use different names, but I have tried to stick with fairly standard terminology that conveys the meaning of the type of work associated with each phase. In short, the SDLC phase names and descriptions presented in this book serve as a foundational approach to developing an SDLC framework, and also serve as a model for understanding other corporate or industry standard SDLC approaches.

Also, an IT organization may employ some standard forms and templates that aid in the proper execution of work described in various SDLC processes. Though not necessary, standardized forms and templates help simplify and improve the consistency and quality of work performed by project team members.

In writing this book, I collaborated with the principals at Documentation Consultants, who provide reusable SDLC forms and templates for their customers.[101] They offer more than 100 standard forms and templates supporting the PMO and waterfall and Agile SDLC practices. Our collaboration primarily focused on developing standard forms and templates for project management and Agile practices. However, their work also provided the foundations for my book titled *Tools and Templates*.

BUSINESS-ORIENTED DEVELOPMENT LIFE CYCLES

The title of this section is not to be confused with the concept of business development. Business development is the set of practices an organization employs to identify and capture new business opportunities to generate growth and revenues. Instead, what I am talking about here is the notion that every business has multiple life cycle development processes. When a new business strategy impacts forced change on the business, the affected areas must go through a transformation process.

The concept of an ECM was introduced in Chapter 1, and I will explore that subject in much greater detail in Chapter 10. For now, it's important to think about enterprise change effects on different components of the business, such as its people, processes, customers, products, and services (e.g., solutions), information, and technologies. Collectively, these components are referred to as the *enablers of the business*. (See Figure 20.)

By merging the life cycle phases associated with the standard systems development life cycle model with the enablers of the business, we can create a matrix view of the scope of change that may be required to support a strategic change initiative across the enterprise.

The main point to be made in this chapter is that IT systems do not alone improve the business. It is the other way around. The organization must change to achieve the objectives of the new strategic business model, and IT systems evolve in lockstep to support and improve the new business processes and organizational changes across the enablers of the business. In short, they go together, as the IT systems must support desired changes across the enablers of the business.

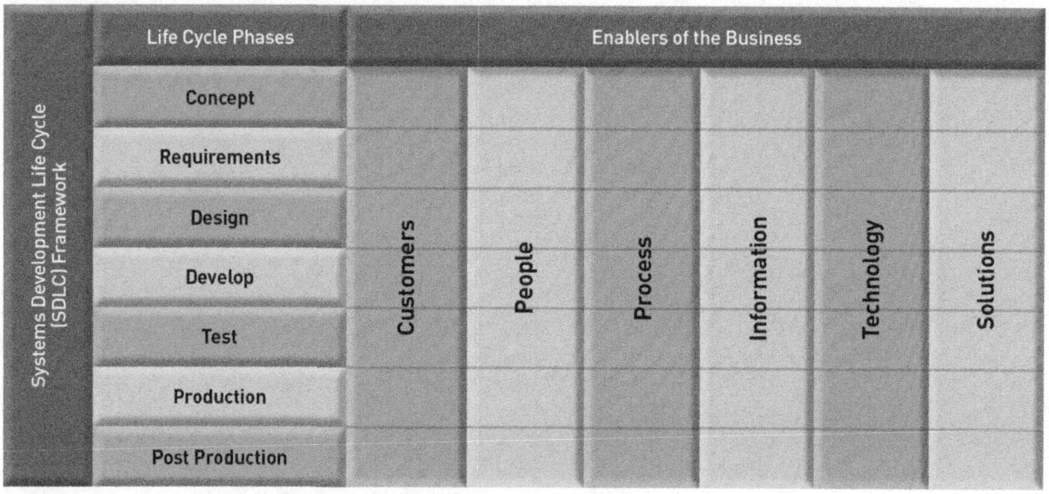

Figure 20. Business-oriented Life Cycle Development Model

Going back to the first section of this chapter, discussing the "system" variant of the SDLC acronym, it should be clear that businesses represent one of the most complex forms of systems that IT can support. While we are familiar with using the term "IT system" to represent a single business application, an enterprise application is more often a solution consisting of multiple business information systems. These systems are integrated to support cross-functional business processes and information flows.

CHAPTER SUMMARY

In this chapter, you have learned three alternative names for the SDLC acronym. SDLC frameworks organize processes by relating them to the type of work performed. While the visual models of an SDLC can imply a flow of work, it's a dangerous practice to think such displays accurately depict the work required at a granular level. Instead, the project team needs to go to the process and task descriptions to discern the dependencies between necessary work tasks.

In this chapter, I have discussed different SDLC frameworks, including those that employ multiple process paths or hybrid SDLC models. The trick of any SDLC framework is to incorporate the best of traditional and Agile practices and make the information available in a manner that allows each project team to evaluate and select the approaches that are most appropriate to their needs, and those of the customer.

Also discussed was the value of implementing hierarchical SDLC framework structures to organize processes by type of work performed or deliverables. Such structures facilitate management and reuse of the organizations standard processes, methods, tools and templates.

The chapter concludes with a discussion on business-oriented development life cycles. Strategic plans drive change throughout an organization, and the impacted areas go through a life cycle set of processes to manage the change. At a conceptual level, the life cycle processes look like the processes associated with the traditional phase based SDLC model. The relevance to IT is that business life cycle changes impact the IT systems that support them.

In conclusion, as funds and resources are allocated to support the changes associated with a new strategic plan, the systems under change may require changes to any or all of the enablers of the business. Moreover, in this context, the term "system" becomes much more broadly defined to imply that a business also operates as a set of complex systems.

09.
SDLCs Are Not Linear

This chapter presents evidence that, both historically and in modern practice, most SDLCs do not follow a truly linear sequential workflow. Though, the linear-sequential profile still persists as representative of the traditional SDLC model, a better and more accurate description of the traditional model is a plan-driven approach that emphasizes planning, requirements analysis and design before starting development related activities.

CHAPTER OBJECTIVES

At the end of this chapter, readers should be able to discuss the following:

- Explain why waterfall approaches cannot be entirely linear and sequential.
- Explain why project management techniques employed in large-scale and complex manufacturing and construction projects influenced the establishment of early waterfall models in software development.
- Explain why evolving IT methods and tools, along with ever-changing business and user requirements, drove the need for more flexible development approaches.
- Explain why modern IT-based systems development practices implemented nonlinear development models with overlapping and iterative phases or stages of work.

AUTHOR'S INTRODUCTION

In the previous chapter, I noted that the linear-sequential model employed in the traditional waterfall model has fallen out of favor. This chapter explains why SDLCs do not work well when implemented as a strictly linear and sequential approach to development. However, I'll also explain how modern SDLC processes overlap, and the work associated with each process is repeated iteratively across multiple development cycles.

I also mentioned that the visual 'waterfall" type display for the traditional SDLC model help set up the errant view that the model promoted a linear-sequential approach to development. As I'll note in this chapter, history shows that this was not the understanding of at least some early users of the traditional model.

In actual practice, a better description for the traditional model is "*plan driven*". The term "*heavyweight*" is often applied to traditional SDLC approaches that are plan driven. This is due to the significant amount of upfront planning, requirements analysis, and design activities that precede development, testing and deployment. In fact, under the plan driven model, there is no reason why the development team cannot employ iterative and incremental development practices. In such cases, the primary difference between the traditional model and agile is the amount of planning and analysis performed before starting the iterative development cycles.

Project management practices are typically associated with the plan driven SDLC model, as this approach seeks to minimize risks through rigorous project planning and use of repeatable processes, with the goal of obtaining predictable results. The plan driven approach also has greater overhead in the form of documentation requirements and third-party quality reviews.

Agile proponents note the traditional practices are inefficient, make debugging extremely difficult and time consuming, and that the linear-sequential processes do not support real-time changes in customer requirements or priorities. All of these arguments are valid when taken to the extremes. Still, as noted in the CollabNet VersionOne Annual State of Agile Reports, more than 50% of the respondents indicated that fewer than 50% of their projects are following agile practices.[102] Statistically, chances are still strong that your project is using traditional SDLC practices. Also, there is no reason why a hybrid approach can't combine elements of the traditional model with agile practices.

The notion that the traditional model must follow linear-development practices comes from the practice of breaking SDLC processes into distinct phases of work. If you were taught a similar viewpoint, I would ask that you reconsider how you view the traditional SDLC phases. Instead of viewing the phases as a strict workflow, view the phases as categories of related types of work. The reason for doing so is that a mature SDLC framework must have a way to manage its SDLC process information, especially its engineering and other non-agile practices, and the phase-based strategy works well for organizing specific types of work.

Moreover, even the traditional SDLC model does not implement SDLC phases in a truly linear sequential fashion. To make my point, this chapter includes a different visual model for the traditional waterfall approach, demonstrating how SDLC processes play out as a series of overlapping phases of work. My goal is not to create a new standard SDLC model or visual aid for the traditional model. Instead, the visual display of overlapping processes helps demonstrate the real-life flow of SDLC work, which is not conducted in a strict linear-sequential flow.

I want to emphasize again that workflow is a matter of dependencies between processes and activities, and the complexity and permutations of those interactions make it nearly impossible to create a generalized visual model at a task level. This visual representation is only a model, and every SDLC-based project plays out differently based on the unique scope of work and needs of each customer.

Dependencies between the various life cycle processes drive the work of a project. Therefore, program and project should expect variations in work across each SDLC. This chapter is relatively short but is ultimately critical to understanding the SDLC hybrid model concepts presented in later chapters, so enjoy!

THE WRONG DEVELOPMENT MODEL

One of the primary issues commonly cited with implementing a traditional waterfall based SDLC methodology is the sequential and linear application of SDLC processes. The problems with this approach include:

- **The need to gather all requirements before starting development lengthened each project.**
- **Sequential execution of SDLC phases and processes further extend development life cycles.**
- **Bugs are difficult to diagnose and fix since testing occurs after development is complete, making it more challenging to figure out which software modules or sections of code are creating the bugs showing up in other areas of the code.**
- **A strict linear-sequential development process does not support modifying the code midstream to support changes in requirements.**
- **The process does not support evolutionary development based on concurrent feedback**

The traditional waterfall approach to software and systems development came from the principles applied in construction, utilities, and large-scale, custom manufacturing operations, where requirements are well defined, and midstream changes would be too costly if not impossible to implement. In such cases, a strictly linear sequence of gathering requirements, conducting requirements analysis, and designing the final product completely before starting any development makes absolute sense.

For example, it is difficult to imagine adding 10 new floors in a high-rise building or adding a new midsection within a ship midway through construction. Could these things be done? Sure, but not without tremendous cost and schedule overruns. In contrast, software development teams need to support changing requirements and an evolving understanding of capabilities possible within a new application.

Some Early Developers Knew Better

Not all early software development projects followed a strict sequential implementation of SDLC processes. As an example, Herbert Bennington notes in the foreword to a paper first published in 1987, "Production of Large Computer Programs," that they employed prototypes, iterative development, and use of enabling software development tools back in the mid-1950s under the Sage Program.[103] Furthermore, he states that the SAGE program was largely successful because they understood that their initial designs didn't always work out and had to be modified several times to get right. Although he didn't use these modern terms, iterative development and testing cycles were crucial elements to their success.

The Sage Program was not the only program using iterative development in the 1950s. In correspondence with Craig Larman and Victor R. Basili, Gerald M. Weinberg made the following comments:

> **"We were doing incremental development as early as 1957, in Los Angeles, under the direction of Bernie Dimsdale [at IBM's Service Bureau Corporation]. He was a colleague of John von Neumann, so perhaps he learned it there or assumed it as totally natural. I do remember Herb Jacobs (primarily, though we all participated) developing a large simulation for Motorola, where the technique used was, as far as I can tell, indistinguishable from XP."[104]**

I reached out to Mr. Weinberg to see if he could further elaborate on this topic. He sent me back an email with the following comments that I found very enlightening:

> **"The quote you cite is about as good as I can do. One thing I might add, though, is the Sage [Program] work was going on at the time I describe, and we had some interaction with Sage people, so ideas were exchanged.**
>
> **I think good ideas don't just come up once, but many times. Aristotle knew that. He said, thousands of years ago, "It is not once nor twice but times without number that the same ideas make their appearance in the world."**
>
> **In other words, it's not of great interest who was first. What I find most interesting is how and why great ideas are lost and have to be rediscovered."**

DIVERGING DEVELOPMENT STRATEGIES

The techniques employed under the Sage Program, according to Bennington, were developed by MIT's Lincoln Laboratory and continued to be used at the Systems Development Corporation (SDC) for another 15 years. In short, prototyping and iterative development were very much part of the SAGE, SDC, IBM, Motorola, and other software programming projects going back as far as the mid-1950s, involving many thousands of software developers.

Military, ISO, and IEEE Standards

Nevertheless, as late as 1988, the programming community at large practiced software and systems development as a liner and sequential set of processes. The use of the waterfall approach was reinforced with the introduction of several Department of Defense (DoD) standards, including:

- Military Standard 490A—Specification practices, published June 4, 1985
- Department of Defense Standard 2167 (DOD-STD-2167) - Defense Systems Software Development standard, published on June 4, 1985.
- Department of Defense Standard 2167 (DOD-STD-2168) - Defense System Software Quality Program, published on April 29, 1988.
- Department of Defense Standard 2167A (DoD-STD-2167A) - Defense Systems Software Development, published on February 29, 1988. [105]
- Military-Standard-498 (a.k.a. MIL-STD-498) Software Development and Documentation, published on November 8, 1994.
- ISO/IEC/IEEE 12207 Systems and software engineering—Software life cycle processes, published originally in July 1995, last updated in November 2017.

Military Standard 490A outlined the required formats and contents of specifications for program related CIs (e.g., hardware, software, or system consisting of both), processes, and materials for systems, equipment, software, and other end items. This standard promoted a documentation intensive life cycle process to design, build, test, and deploy software and systems.

DOD-STD-2167 and **DOD-STD-2167A** established uniform requirements across the life cycle of software development. These two standards did not explicitly define the execution of SDLC processes as a linear-sequential or waterfall-based approach. However, conformance with the detailed documentation requirements would force the issue due to the substantial work up front to document software requirements analysis and design.

DOD-STD-2168 was written as a companion to the DOD-STD 2167 standards and established requirements for the development, documentation, and implementation of a software quality program. The purpose of this standard was to encourage a standard set of practices to help achieve, audit, and maintain the level of quality in software and software quality documentation.

MIL-STD-498 established a uniform approach to software development and documentation requirements.[106] At 516 pages, Mil-STD-498 provided a comprehensive description of SDLC processes and served as a baseline for all ISO, IEEE, and other systems development standards that came after it. On May 27, 1998, the DoD formally moved away from developing standards and adopted the ISO/IEC/IEEE 12207 for the development of DoD software systems.

DID YOU KNOW?

MIL-STD-498 was designed to scale up to support the development of any large-scale systems, such as computer systems, defense systems, aircraft, battleships, canals, dams, factories, satellites, and submarines.

MIL-STD-498 specifically promoted the use of iterative development techniques, and also noted that the listing of defined activities within the standard did not imply a specific ordering in their use. The standard went further to note that waterfall practices force time-based dependencies plus overly formal and bureaucratic requirements lead to budget and cost overruns.

ISO/IEC/IEEE 12207 defines standards for software life cycle processes and terminology. Specifically, the international standard defines processes for the acquisition of a system that contains software, a standalone software product, and software services.

Originally, the standard evolved from MIL-STD-498. However, the standard has been updated several times since its inception, with ISO/IEC/IEEE 12207:2017 as the latest version.

On the other hand, MIL-STD-498 imposed a significant documentation burden in the form of data item descriptions (DIDs). [107] The DIDs included a set of 22 plans categorized within 7 areas of work in lieu of a standard SDLC process (see Table 13). Development projects using the MILSTD-498 standard were free to choose the DIDs that best supported their needs.

Still, can you see how Agile developers might be concerned about the amount of documentation forced on them through these early standards? [108]

Table 13. Data Item Descriptions (DIDs)

Plans	Software Development Plan (SDP)
	Software Installation Plan (SIP)
	Software Transition Plan (STP)
Concept/requirements	Operational Concept Description (OCD)
	System/Subsystem Specification (SSS)
	Software Requirements Specification (SRS)
	Interface Requirements Specification (IRS)
Design	System/Subsystem Design Description (SSDD)
	Software Design Description (SDD)
	Database Design Description (DBDD)
	Interface Design Description (IDD)
Qualification/test products	Software Test Plan (STP)
	Software Test Description (STD)
	Software Test Report (STR)
User/operator manuals	Software User Manual (SUM)
	Software Input/Output Manual (SIOM)

	Software Center Operator Manual (SCOM)
	Computer Operation Manual (COM)
Support manuals	Computer Programming Manual (CPM)
	Firmware Support Manual (FSM)
Software	Software Product Specification (SPS)
	Software Version Description (SVD)

Table 13. Data Item Descriptions (DIDs)

EVOLVING METHODS AND TOOLS

As noted previously, during the earliest days of software and applications development, the only model for developing complex products came out of large-scale manufacturing and construction projects. Outside of a few enlightened programs, such as the SAGE Program, there wasn't an industry standard model for building expensive, complex products that also were subject to an evolving understanding of requirements and external change drivers.

The IT development paradigm is different than those found in construction and manufacturing. Unlike the physical world, where architectural drawings can help customers imagine a product before construction, software architecture and design documents are logical representations and difficult to casually understand and interpret. On the other hand, in the IT industry end users can view mockup screens and relatively inexpensive prototypes before establishing their final requirements. The prototypes and iterative releases of a developing product provide users and customers with new insights into the potential capabilities of the solution, allowing the software products to evolve with the requirements.

Iterative development is important because competition, business requirements, and new opportunities drive organizational change, and supporting business systems need to evolve to support the business as it evolves. As a result, the IT industry required more flexibility in the design and creation of software. Fortunately, the IT industry responded, creating integrated tools that support UI design, modeling, prototyping, and rapid and iterative development capabilities. In other words, newer generations of software development tools better supported rapid and customer-centric development activities, supporting the principles and values advocated by Agile practitioners.

No Two SDLC Models or Instances are Alike

Each organization must ultimately decide how they organize their SDLC frameworks and the processes they require. Different organizations may use different names for the phases or stages of their standard life cycle models. They may even have a different number of phases, based upon the granularity of their systems development framework. It's possible not to use a phase-based model at all and instead create a framework around supported methods and tools. There is no right or wrong in this, as long as it works for the organization, and they implement effective programs to document, implement, and improve their set of best development practices.

Also, SDLC frameworks that include multiple systems development models may have different processes, activities, tasks, and dependencies associated with each. Such SDLC frameworks allow each program and project flexibility to tailor their project approach to support variations in their scope of work, complexity, risks, and budgets. When organizations take this approach, the SDLC framework needs to provide guidance on which approaches work best under different conditions.

Start Project Planning Early

Some project management disciplines promote hiring the project manager with the approval of a project's charter, and the PMs then put the project team together and start the project planning effort. I couldn't disagree with this strategy more. If the project team waits until the project charter is approved, they are already behind the eight ball in terms of figuring out how they should run their projects.

In an ideal situation, the project manager and senior technical leads should be involved in project planning during a product's conceptualizing phase so they can provide input to the approving customers or executives. That is the only practical time they have to influence the scope, budget, schedule, and quality constraints specified in the project charter. Business cases and ROI analyses are of no value without sufficient planning to assess the entire scope of work necessary to define, develop, and support the products.

It is also ideal to approach the response to an RFP as the beginning of project planning. How else can the contractors understand if the prospective business is profitable under the constraints dictated by the customers? It's an even better situation when the contractors' project planning process starts with the initial RFI, so their responses better inform the customers who need to assess the vendor information against their constraints before putting out an RFP.

During product conceptualization, the project team should already start working with their PMO and executive managers to assess resource requirements and decide how they should approach development. Project resources, schedule, and client/user access may impact some of these decisions. These project-based modifications determine the extent of overlaps in each of the life cycle phases, and the individual phase durations.

NON-LINEAR SDLC MODEL

Visual representations of IID models typically take a circular form, such as with the Scrum methodology. There's nothing wrong with those views, as such images convey the notion that systems development work is repetitive in nature. The problem I have with most circular representations of agile processes is that they either imply sequential execution of tasks in a repetitive manner, or they are at such a high level of abstraction that the actual analysis and engineering processes are ignored.

Figure 21 solves this problem by depicting the phase based SDLC processes within an agile IID wrapper to indicate similar types of work are performed across multiple development iterations, but not necessarily in a linear-sequential fashion. I like this visual model as it demonstrates how each iteration of development works on a slice of the total scope of work through the traditional set of life cycle engineering and analysis processes.

However, it would be a mistake to think all life cycle processes are required in every iteration. The development team must assess the scope of work for each iteration and execute only those SDLC processes that are required.

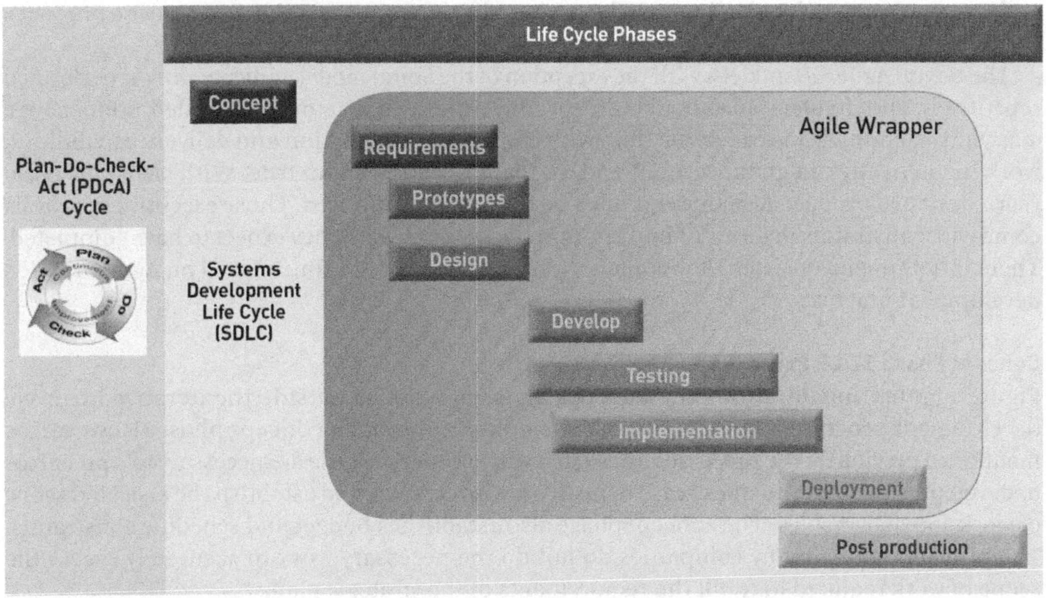

Figure 21. SDLC Phase Overlaps in Context with Project Management Process Groups

The left-hand side of the model shows the interaction of PMLC processes through the standard Shewhart-Deming PDCA or PDSA cycles. This is the discipline that IID practices help enforce. In other words:

- **Plan** - the work required within an iterative development cycle,
- **Do** - the work,
- **Check** - to confirm the expected work was completed in conformance with the requirements,
- **Act** - to address any risks or issues, and determine the work for the next iteration

Over the course of a product's life, it is possible to initiate multiple projects spanning different pieces of work. New projects are spawned every time executive management approves new work and funding to support new product requirements. The concurrent execution of project management processes helps ensure the team stays focused on staying within the customer approved boundaries of scope, budget, schedules, quality, and resources.

Project Management across the SDLC

Note that the Figure 23 graphic also includes standard project management process groups, as defined by the Project Management Institute, listed in a column to the left of the SDLC.[109] The project management processes do not map directly to any of the phases in the SDLC. The intention is to demonstrate project management processes play a critical role in managing SDLC based processes, but also operate entirely independently and separate from SDLC processes.

For example, a project may be initiated to develop the concept for a project. Another project might be initiated separately to conduct an initial analysis and documentation of requirements and designs. Additional projects can be initiated for each new release of the product. Each new project is managed separately following the organization's project management life cycle (PMLC) processes.

Moreover, depending on how the product is approved and funded over its life cycle, and also on the complexity and size of a system or product, customers and executive management may fund multiple projects to oversee different phases of a system's evolution. In such scenarios, each new project team plans and executes the project management processes across a specific and approved scope of work. There can be any number of projects established across a product's or system's life cycle, to support future enhancement requests.

The typical Agile/IID models, with the exception of the Spiral model, indicate work is performed repetitively. The circular visual models do not convey the idea that projects, by definition, have a start and an ending. One can argue that, with continuous integration and delivery capabilities, work theoretically can go on without end. In reality, that's seldom true. With each new fiscal year, continued product development must be funded and approved. Those executive approvals come with constraints in terms of budgets and the scope of work they expect to have completed. There is nothing inconsistent about having project-based constraints placed on Agile-oriented development practices.

Concept Phase SDLC Processes

Figure 23 intentionally shows the concept phase of work as outside the iterative life cycle development processes. Some Agile models do not promote a concept phase. However, as mentioned previously, the processes associated with a concept phase are necessary as a precursor to developing the basic business case to justify the investment and establish the expected scope of work and deliverables. The concept phase also establishes budget and schedule constraints. The problem is that many companies do not do the necessary work to accurately access the scope of work required to fulfill the requirements of the strategic plan.

A CONOPS investigation, which is an assessment of end user requirements to support a new business strategy, can logically fit in either the concept phase or the requirements phase. However, the high-level requirements identified during a CONOPS investigation help establish the scope of work identified in the program or project charter. Therefore, the earlier it is completed, the better. In my book titled *Frameworks for Agility*, the CONOPS process is conducted in the concept Phase.

Parallel SDLC Processes

There is another approach to iterative development that is more difficult to convey visually, and that is the execution of parallel SDLC processes. This is an important concept as it fits the continuous integration and delivery models. CI/CD capabilities require process integration and automation to synch the activities and use lean-Agile practices to minimize wasted effort and disordered work.

Lean-Agile practices strive to eliminate waste in the development processes by reducing work in progress. To make such a system work, each phase of product development occurs in lockstep. In other words, work proceeds through the life cycle development processes at roughly the same rate.

SDLC processes are but one of many organizational processes, and SDLC process automation is simply another form of business process automation. Modern technology and IDEs link and order the execution of SDLC processes. The processes follow the standard phase-based model, but at a more granular level of detail. I discuss the topic of SDLC process integration and automation in much greater detail in my *DevOps* book.

For larger, enterprise class systems, it would be a bit more efficient to have a single feature developed in detail before starting on development of the next feature, in parallel, but at different stages. Also, multiple teams might be working on the same product at the same time. Therefore, I want to make the point that multiple development activities can go on at the same time, by different teams who are executing different types of work on their own iterative schedules.

CHAPTER SUMMARY

In this chapter, we have discussed early and successful implementations of prototyping, iterative development, and use of tools to simplify development, as evidenced by MIT's Lincoln Laboratory work on the development of the Semi-Automatic Ground Environment (SAGE) system, and other early adopters. Unfortunately, other IT development organizations stayed with a much more traditional approach to developing expensive and complex systems that enforced the use of phase based SDLC processes, executed linearly and sequentially. These practices were encapsulated in several early systems development standards, further enforcing bad habits.

Readers have learned that modeling and software development tools advanced to support the unique nature of the software development industry. These new methods and tools make it easier to create prototypes, and for developers to iteratively develop software and release more frequently, and to support changing or evolving business and customer requirements.

The chapter presents a visual model that shows how SDLC processes play out repetitively in the modern IID development paradigm. The SDLC phases do not imply a specific process sequencing, as the processes tend to overlap and often repeat in each iteration. Rather, the dependencies between processes drive project workflow and schedules.

Readers have learned that no two standard SDLC models are the same, as each organization's resources, organizational structures, methods, tools, and best practices differ. Therefore, each IT organization needs to establish, mature, and maintain a set of SDLC processes in a form that suits their specific business needs.

A mature corporate SDLC should be flexible, offering alternative development strategies. Each project team must tailor the organization's SDLC processes to incorporate the activities they require to best support their customer's requirements.

10.
Enterprise Change Model

For readers who want to understand how an ECM helps assess the impact of strategic change initiatives and help guide cross-organizational and cross-functional business process assessments that ultimately affect the business systems supporting the business.

CHAPTER OBJECTIVES

At the end of this chapter, the student should be able to discuss the following:

- Explain how strategic change initiatives affect business systems development.
- What are business enablers, and why do they matter?
- Explain how a life cycle development framework can be built with the goal to help manage change across the enterprise.
- Describe how to construct a conceptual ECM framework as a matrix with business enablers and life cycle phases.
- List and describe the five key concepts that underlie the ECM.
- Discuss and provide examples of how changes in one business enabler can impact others, forming dependencies between the enablers that are not always in sync across their respective life cycles.
- Describe the type of information captured and maintained on(in?) the processes and activities associated with each life cycle phase for each affected enabler.
- Describe how dependencies between the enablers drive the schedules in an enterprise or strategic change initiative.

AUTHOR'S INTRODUCTION

We are now at a point where we can discuss how the organization can employ life cycle processes to help manage change across the enterprise. The topic is important because strategic or enterprise change initiatives affect organizational resources and business processes, which in turn drive requirements for the business systems that support them.

The material presented in this chapter could very well fit within a book on management consulting practices, with a focus on the effects of strategic change initiatives on cross-functional and cross-organizational business processes. These notions are not novel. For example, in the early 1990s, I worked within a Business Consulting Service Development organization for an IT based professional services division within a large multi-national firm. Though we described our ECM a bit differently, the basic concepts promoted in this chapter are similar to those we used to help assess the impact of strategic change on critical business processes.

In the previous chapter, I briefly introduced a business development life cycle model and noted the concept of life cycle phases applies to any number of business enablers affected by strategic change initiatives or portfolio investment decisions. Now we'll put these concepts together, in a matrix format, to create a framework to organize life cycle processes that support change across the enterprise. The value at looking at an enterprise from this perspective is to understand both the complexities and interrelationships involved in a strategic change initiative.

BUSINESS AS A SYSTEM

When defining a standard SDLC framework for the corporation, the systems under development can be much more complex and involved than simply developing software or information systems. In support of business strategies, organizations may also need to effect change in other critical enablers of the business such as organizational structures, skills, resources, information, technology, equipment, and supporting business processes.

In a business context, IT provides the grease that helps the organization operate efficiently. And no amount of grease can improve the functionality of poorly designed mechanical or business systems.

We have noted several times that life cycle processes are about managing change, and such business transformations must support the strategic goals and objectives of the organization. When business executives look at the things they need to change within the organization to be competitive or profitable, or to develop the right set of products and services for their customers, they must look at all the corporate or organizational assets that help the enterprise achieve its business strategy.

At a high level, organizational assets include customers, people, processes, technology, information, and partners. At a lower level, though not meant to imply an all-inclusive list, organizational assets include organizational structures, culture, skills, roles and responsibilities, customers, data, equipment, operations, delivery, support services, sales channels, and supply chain & delivery partners. Collectively, these corporate assets can be thought of as the business enablers.

Enablers of the Business

Let's look at some of these business enablers more closely.

CUSTOMERS

Customers have "needs" that organizations address via the products and solutions they offer. Before considering change requirements across the other enablers, we must understand that fulfillment of customer needs is the primary driver for all other change initiatives. Yes, profitability, cost management, and a host of other organizational needs are important.

Paying customers are the folks who ultimately enable the commercial business. Without customers, the business fails. Similarly, citizens are the customers of government agencies, and their needs drive the mandates of those agencies. Nonprofits are interesting in that their customers include both the benefactors and beneficiaries; both have "needs" that drive change initiatives within the nonprofit.

Organizations may provide different products and services for different segments of their larger target market. The need to ensure survival, or to achieve growth and profitability objectives, drive organizational change initiatives. In such cases, customers drive the need for new products and services.

Note also that customers can be internal organizations that require support from other internal organizations. For example, the IT department serves the needs of the other departments by providing network access, email services, application support, and development of business systems that facilitate the organization's critical business processes.

SOLUTIONS

IT systems evolve to support the identification of customers and their needs, order delivery/ fulfillment, and the organization's solution offerings to customers.

Solutions are the final customer-oriented deliverables provided by the organization. These can take the form of products, services, information, or other outcomes. Solutions are created to provide fulfillment to customer needs, and for that reason are directly linked to the other business enablers.

To develop a customer-oriented solution, we need to identify who our customers are, discover what pains and needs they have and evolve a solution to deliver the capabilities they require at a price they can afford. The same definition for solutions is true whether we are describing an organization that supports an internal customer, supports an external commercial customer, or supports the provisioning of government and nonprofit services.

IT SYSTEMS

IT systems evolve to support organizational processes as well as systems involved in the life cycle processes required to define, build and deploy solutions offered by the organization.

IT systems evolve to help manage information related to the people who work in or for the business. For example, the organization may deploy an HR portal to support its internal employee information distribution requirements. Other systems support provisioning of employee benefits, making payroll, recording timekeeping, and making payments on employee expenses.

PEOPLE

People as business enablers include the types of resources, skills, and experiences necessary to execute the corporate strategy and to support the development and delivery of products and services across their life cycles. The People business enabler includes management and labor available to the business both within and external to the organization. For example, the People enabler includes the organization's support staff such as administration or HR, external partners who provide skills and resources, and other systems support staff.

PROCESSES

Processes enable the business by providing efficient, practical, proven, and repeatable methods to carry out the work that is necessary to support the corporate strategy. Some processes are standalone. However, most processes are integrated with other business processed to properly execute work or guide the flow of information across the organization.

From an IT perspective, enterprise business systems evolve to support the efficient and accurate execution of the organization's business processes and flow of information. Applications supporting business processes can be custom-built. However, there are numerous COTS enterprise software systems that can be tailored to meet the needs of most organizations in most industries. A word of caution, a COTS-based solution typically needs tailoring, if for no other reason than to support unique needs of the business and to implement processes that make the business efficient and competitive in their markets.

DATA

Data includes the raw facts or information about the business and its customers, partners, products, and services. There are two primary data types: structured and unstructured. Structured data includes the basic alphanumeric strings and numbers used in a relational database or worksheet and easily classified as to their contextual use. For example, contact information such as name, address, city, state, and zip code are all forms of structured data. We know what these terms mean, and the expected formats of the data are related to those terms.

In contrast, and as the name applies, unstructured data does not have such a structure. Unstructured data can include unformatted text files, video or audio files, emails, text messages, image files and data captured from electronic sensors.

From an IT perspective, data needs to be collected, organized, transferred, manipulated, reported, stored, managed, and made usefully available. Structured data is handled fairly readily in relational databases and worksheets as records in tabular formats. Unstructured data requires the attachment of metadata (data about other data) to define what the data means and how it can be used by individuals or other software applications and systems.

In raw form, both structured and unstructured data is managed within a database management system (DBMS) and file-based directories of various types. Database experts work with people inside and outside of the organization to determine the data requirements necessary to support other enablers of the business. Data is used and displayed within software applications and other systems and turned into useful information via reporting and other business intelligence tools.

When enterprise data is managed and stored centrally, the repository is referred to as a data warehouse. In the past, there were technical reasons for managing data external to the large transactional business applications that were tuned to support high volume processing of data. By contrast, business intelligence tools that performed online analytical processing against a system tuned for transaction-oriented processing could bring the system to its knees from a performance perspective. So, it was better to move the data offline to a standalone database management system. The tools used to move the data and to put the data into the correct formats for the receiving repositories are called extract, load and transform (ELT) tools.

BIG DATA

Modern technologies have reduced the performance issue of querying data on transaction-oriented applications. However, data warehouses live on as central repositories supporting Big Data initiatives. Relatively cheap data storage devices combined with the enormous growth of interconnected devices and applications make possible the ability to find correlations and subtle patterns in data coming from distributed and disparate data sources.

Some IT analysts like to communicate the characteristics of Big Data with a vocabulary of words starting with the letter "V." Today, some analysts have greatly extended the list of descriptive words starting with the letter "V" to as many as forty-two descriptive terms.[110]

Here's a funny story. When I asked permission to cite Tom Shafer's article titled "*42 V's of Big Data and Data Science*," I was informed that the information, though factual, was created as a tongue-in-cheek poke at the "V's of Big Data" for an April Fools post. The point being, we can all get carried away with being creative in the use of terminology, and yet fail to communicate the total scope of the subject accurately.

Since this is not a book on Big Data, here is a sample of six characteristics commonly found in such lists on Big Data that also happen to start with the letter V.

- **Value**—Uncovering relationships and patterns that are useful but hidden and difficult to find in the sheer volume of data available today
- **Variety**—The ability to connect disparate data sources and exchange information between systems using different data types and formats
- **Velocity**—The speed at which new data becomes available and also the speed at which it becomes outdated
- **Veracity**—The degree to which data is accurate, precise, and trusted
- **Visibility**—The ability to interact with the data to see relationships and patterns

- **Volume—The sheer volume of data and continued exponential growth driven by cheap storage and the massive number of interconnected devices and systems**

INFORMATION

Information is data made available in useful context and forms. Most dictionaries define *information* as knowledge or intelligence that is learned or discovered. True, but to what point? Certainly, intellectual capital exists in the heads of the organization's expert practitioners. Moreover, an organization may employ knowledge management systems, information portals, and other repositories to capture and maintain their critical knowledge assets. However, that's not the complete story on information.

Information is further defined as data put into useful context to address some business or personal need. In other words, data has little use without understanding its meaning or what it is telling us in the context of its use. Data put in context tells us something useful. Turning data into useful information is the primary objective of Big Data.

From an IT perspective, converting data into information is the function of big data, business intelligence, data mining, and reporting systems. Some business applications and other systems allow input and display of information within fields. The applications may also have limited reporting and analysis capabilities. However, it's also not unusual to have data moved into a reporting or business intelligence tool for more sophisticated analysis.

TECHNOLOGY

Technology is the application of technical or scientific knowledge in support of the business strategy, goals, and objectives. Broadly, this term encompasses the development or procurement and application of all intellectual property, methods, and techniques that help the organization achieve its strategic goals and objectives. Technology supports all the business enablers by offering new capabilities or enhanced performance and efficiencies.

Information technology (IT) is one category of the technology business enabler. Information technologies are constantly evolving and collectively serve as a critical and constant driver in supporting business strategies and related organizational change initiatives. IT also supports other technology-based business enablers.

For example, new equipment can be automated through IT software and other IT components. Information technology can enhance the capabilities of Customer-oriented Solutions. Business systems support the information needs of Customers, Partners, People, and Processes. I've already discussed how IT-based systems and tools convert Data into contextually useful Information.

EQUIPMENT

Equipment includes any physical apparatus, device, machine, tool, or transport device or vehicle used by the business in support of its business strategies and processes. In an IT context, we think of computing equipment, networks, and storage systems; but this category can also include things like manufacturing and transportation equipment, tools, fixtures, and virtually any tangible property outside of buildings, structures, or property, that is used to support the business. We can also think of Equipment as being a form of Technology realized in a physical form.

As with Technology, software applications and IT systems and components extend the capabilities, performance, and efficiencies of Equipment.

ORGANIZATIONS

Organizations, as business enablers, are a collective structuring of human and other resources to accomplish the business strategy. Business entities typically require many organizations, each aligned to support a specific business function, or aligned as a matrix of cross-cutting resources.

Functional organizations are hierarchical in structure, support specialization of skills and roles, and provide centralized authority. Functional organizations are enduring, implemented to develop human resources, skills, and processes, and best suited to operations-oriented work where change is minimal and controlled.

DID YOU KNOW?

Organization change is stressful. Change affects the morale and culture of people, which can have a negative impact on any organizational change initiative.

ORGANIZATIONAL CHANGE MANAGEMENT

Organization change management (OCM) is a framework of processes to help an organization manage the negative effects of change on its people.

OCM specialists communicate the vision and reasons for the change, help set expectations and sell the benefits of the change, and reward people for their support. OCM specialists also educate employees on how their work must change and help establish metrics to gauge success when implementing the proposed changes.

In contrast, a matrix organization includes a collection of cross-functional resources that contain all the types of skills and resources deemed necessary to perform work in a highly dynamic environment, such as that within project-oriented work. Given that projects are temporary and allocated over a specified duration, resources assigned to a project team may come from the business' functional organizations on a temporary or part-time basis.

IT supports organizations primarily through business systems that support cross-organizational and cross-functional business processes. IT systems help improve workflows and data flow across the organization, usually in support of the organization's business processes.

PARTNERS

Partners include a wide range of external organizations and people necessary to support the operations of the business and its corporate change initiatives. For example, partners may augment the organization's capabilities in marketing, supply chain management, manufacturing operations, program and project management, professional services, consulting, and solution delivery.

Businesses enter into partnerships when the external organizations are perceived to have cost, skills, experience, and delivery advantages. In other words, partners provide services that would otherwise be too expensive or take too long for the company to develop internally within the time available to support the strategic plan.

IT supports supply chain Partners with information systems that support order entry and coordination of supplier deliveries in alignment with the needs of the manufacturing groups. Other IT systems may help coordinate deliveries and installation of products by third-party delivery Partners.

SALES CHANNELS

Sales Channels are separate from Partners, as the relationship here is one where the Channel Partner receives compensation from the sales of the product to the organization's Customers, while other Partners tend to make their income from selling products and services directly to the business organization. Channel Partners can include outsourced sales organizations, resellers, distributors, wholesalers, and any other form of external sales and marketing organization.

Another source of sales channels are companies that purchase products from the manufacturer with the goal to produce and sell derivative products. Such relationships fall under the terms of an original equipment manufacturer contract agreement.

IT systems help connect processes with those of the provider business through the interoperability of their respective business systems or other means of data transfer. The organization's IT systems send new orders to supply chain partners, take orders from sales channel partners, and provide information on available product and inventories.

A Simplified Business Transformation Model

Each of the identified business enablers is a resource to the business, which is why we previously referred to them as organizational assets. The impact to all business enablers has to be considered during any business transformation initiative, as each enabler plays a role in the ability of the organization to achieve the strategic goals and objectives mandated by the organization's executives.

Also relevant to this discussion, when the organization is evaluating alternative strategic plans or portfolio investment decisions, is the fact that the scope of change identified in each may affect a different subset of business enablers. In other words, every proposed change initiative has a different impact on the enablers of the business.

Business Transformation Impact on Business Systems

The previous section included a simplified model of the business enablers and discussed how strategic or transformational change initiatives could impact one or all of the business enablers.

IT systems are built to support the information and workflow needs of the organization and must evolve to support changes imposed through the execution of a new business strategy.

The enablers of the business create a complex and interrelated system of processes and information flows which the business systems must emulate to support. In that context, one can think of an enterprise business system and other business process support systems as the implementation of a model of how the business should operate.

The relevance to our SDLC discussions is that changes in the business enablers often force the development of new business systems or modifications to existing systems. Though this is a book on system development life cycles, the organization must take a holistic view of the impact a strategic change initiative has on the business enablers and business systems that support them.

Strategic Change Life Cycles

If the organization aspires to have a standard SDLC framework, they might also want to develop a standardized life cycle view of how they plan to organize and manage their enterprise change initiatives. In such cases, the organizations would develop a taxonomy or classification schema to name, organize, and manage their business-oriented life cycle management processes.

Working through a strategic change initiative logically, we can imagine a set of life cycle processes that look similar to the phase-based life cycle processes of the traditional SDLC model.

- **Concept**—Document the high-level capabilities and business justifications in each business enabler to support the revised mission and strategic plan.
- **Requirements**—Document the detailed functional and nonfunctional requirements across each of the affected business enablers.
- **Design**—Conduct as-is/to-be/gap analysis to determine what the future state will need to look like compared to the current state, and the amount of change required in each of the affected business enablers.
- Development—Acquire/build partnerships, equipment, systems, and other necessary components in each of the affected business enablers. Potentially prototype the proposed changes on a limited scale.
- Test—Make sure the changes are effective in an environment that minimizes risks to the organization.
- Deployment—Transition the changes to the organization, including necessary acquisitions and execution of contracts, plus deployment of equipment, process modifications, training, and systems.
- Support/Sustain—measure the performance of the changes against the metrics of the strategic plans and adjust accordingly.

As with the SDLC, ECM phases describe standard processes, activities, tasks, best practices, tools, roles and responsibilities, and deliverables across the life cycle of each business enabler. These relationships are shown more succinctly in later sections of this chapter.

We know the IT organization supports change and operations across all business enablers. However, when we look at a business as a large and very complex system, IT has its biggest impact on the implementation and improvement of the organization's business processes. The processes can change all at once in a business transformation effort—usually as part of a BPR effort or improved incrementally over time—as part of a BPI activity.

This life cycle model is usefully descriptive of how to manage across the business enablers and related business processes. However, we can also apply the concepts of the traditional waterfall and Agile based development strategies to manage enterprise change. In this sense, one can think of BPR as following a waterfall-based approach to making a significant advancement all at once, while continuous process improvement activities involve Agile-like incremental change.

Enterprise change initiatives need to be planned, managed, scheduled, monitored, controlled and reported on, just as we would for traditional software and systems development projects, ever mindful of the dependencies that must be discovered and coordinated across each life cycle. Again, strategic change and portfolio investment decisions often drive development projects across the enablers.

A LIFE CYCLE VIEW OF MANAGING ENTERPRISE CHANGE

Chapter 2 introduced the concept that project management and SDLCs, used in conjunction with each other, help the organization manage the complexity of developing complex IT-based systems and software products. Specifically, SDLC processes help ensure organization-wide use of mature practices to develop complex and unique software products and systems, while PMLC processes help ensure that the organization has mature practices to manage such project-based development activities.

The same concepts work for managing large scale and complex organizational changes associated with implementing a new strategic plan or other large-scale change initiatives. In other words, the organization takes a life cycle view to manage change across the enablers of the business.

Business Enablers

Business enablers are enterprise assets that must be managed and coordinated to optimize corporate efficiencies. Table 22 provides an example of categories of business enablers and underlying corporate assets associated with each enabler.

Categories of Business Enablers and Underlying Corporate Assets

Business Enabler Categories	Focus	Purpose
Customers	The reason the organization exists—to provide products and services that target market customers want	**To analyze market opportunities and attract and keep target market customers** • Target market identification • Value proposition fulfilment • Customer retention programs
People	Finding and developing human resources as assets to support business operations across the enterprise	**Providing necessary skills and human resources** • Employee resources • Contractor resources • Partner resources • Stakeholders
Process	Analyzing, developing, and streamlining collections of activities that help make the business efficient and effective; can be departmental,	**To evaluate business integration needs and performance improvements** • Account management • Accounting • Customer relationship management

Business Enabler Categories	Focus	Purpose
	cross-functional, and cross-organizational. Organizations that practice lean concepts also seek to eliminate non-value added or wasteful activities and tasks that do not directly support the mission of the business or its strategies, goals, and objectives.	• Customer support • Demos, prototypes & workshops • Distribution • Financial management & Controls • Governance, policies, and procedures • Human resource management • Inventory management • ITSM/ ITIL • Lead generation • Lead qualification • Marketing communications • Order management • Partner management • Payroll • Procurement • Product management • Project management • Proposal management • Process maturity/quality management • Sales prospecting • SDLC • Supply chain management • White paper generation
Information	Developing, delivering, analyzing and reporting information plus protection of knowledge assets	**To protect and make available organizational information assets** • Business performance metrics • Business process information • Customer information • Data analysis • Intellectual property management • Knowledge management • Reports and dashboards
Technology	The application of scientific or specialized knowledge as assets across the enterprise	**To implement technology-based improvements, spanning:** • Facilities • Manufacturing equipment • Materials • Methods • Software/IT systems • Transportation equipment • Tools
Solutions	Designing, creating, and delivering customer-centric products and services	**To provide competitive and unique offerings within the target market clients' budget and other constraints** • Products • Services • Other deliverables

Figure 22. Categories of Business Enablers and Underlying Corporate Assets

This model is conceptual, and nothing prevents us from adding to, deleting, or modifying the business enabler categories, or the underlying assets included in each category. For example, we might want to include Financial Capital or Infrastructure as business enabler categories. If the organization's executives choose to add a business enabler for Infrastructure, they might move the Equipment, Facilities and Transportation assets under that category. The point is, the categories need to make sense to the organization, and at a level that is of sufficient detail to document the assets of the business that are subject to change, yet not at a level of detail that is unwieldy or unsustainable.

Every organization has different areas of emphasis. The primary purpose of the model is to provide a guide to think through the areas where a strategic change initiative might impact the organization.

Enterprise Change Model (ECM)

I've noted that business enablers have life cycles, just as with any tangible or intangible asset.

The linear-sequential life cycle model is a standard model in project management. The standard life cycle model by itself doesn't provide any detail on how to build or create things. The lower level processes provide that type of information. However, the structure was and remains useful to conceptualize the type of work that occurs across the life cycle of any organizational asset that needs managed change.

With this in mind, Figure 23 displays a life cycle approach to managing enterprise change as a matrix view of life cycle development phases in context with critical enablers of the business. This model is called an enterprise change model (ECM). The ECM provides a tool that allows an organization to develop a set of processes and best practices for managing change on an enterprise scale.

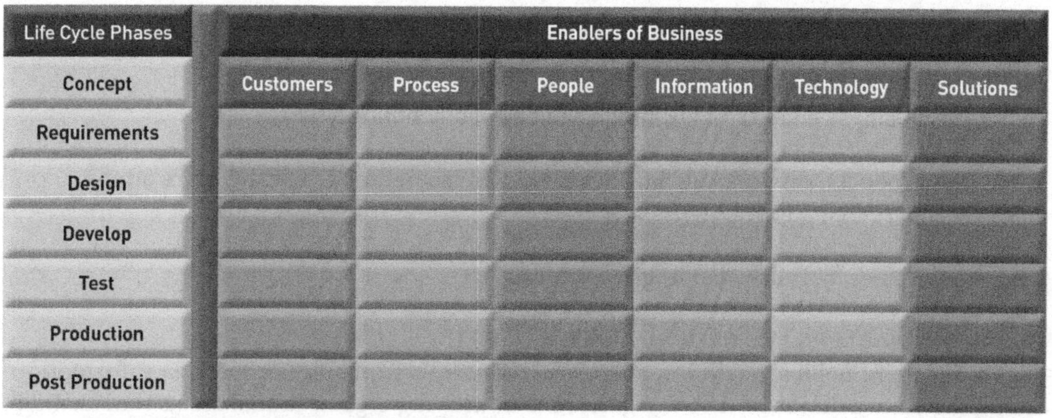

Figure 23. Enterprise Change Model (ECM)

ECM and SDLC Similarities

Looking at Figure 24, it should be clear that there are similarities between SDLC and ECM frameworks. Where an SDLC provides a framework to categorize, organize and manage IT processes, an ECM provides a similar framework for managing other corporate development life cycle processes. The overall solution for an SDLC is a new or modified IT product. The solutions for an ECM are additions and modifications to the corporate assets that fundamentally enable the business.

Each framework starts with processes to flesh out the concepts to see if they are practical to pursue, and what the final result needs to be. A requirements phase explores solution requirements in more depth, in context with the customer and end user needs. A design phase provides processes to analyze architectural and design requirements necessary to deliver on the solution requirements. Development and testing phases create the solutions and ensure that they conform to the requirements. Solutions developed under an SDLC or ECM are deployed, maintained, and finally come to an end of life.

The main point of this comparison is to point out that mature organizations have an advantage through the implementation and maintenance of life cycle development processes. That is true regarding the products and services they create for their customers, the IT systems developed for internal use or for external customers, and for other corporate assets that enable the business.

A MATRIX VIEW OF ENTERPRISE LIFE CYCLE PROCESSES

Where an SDLC is a single-dimensional framework to organize life cycle development processes for software and IT systems, the ECM is a multidimensional framework to organize life cycle processes across all identified enablers of the business. In other words, an ECM applies life cycle phases in context with the business enablers to help the organization succeed in its mission and business. The result, from a visual perspective, is a matrix view of life cycle phases across the identified business enablers.

The model shows how each business enabler has a life cycle that can be affected by strategic change initiatives, often driven by target market customer requirements and related portfolio decisions. Take a moment to review this ECM framework, as I'll build on the ECM concept throughout the remainder of this chapter.

Key Concepts

The ECM is constructed to visualize and support six fundamental concepts:

- Business enablers are organizational assets.
- Target market clients drive portfolio investment decisions.
- Organizational change involves project-based work.
- Customer and business needs assessment is key to scoping change.
- Business enablers have integrated development life cycles.
- Products, services, and businesses evolve; sometimes radically, more often incrementally.

Let's take a closer look at each of these concepts.

BUSINESS ENABLERS ARE ORGANIZATIONAL ASSETS.

The high-level business enabler categories—customers, people, processes, information, technology, and solutions—help categorize the underlying organizational assets that facilitate the functioning of the business. Every business leverages organizational assets across product, service, and business life cycles to provide the solutions their customers want; these organizational assets enable the business. In other words, organizational assets are all business enablers, expressed at a more granular level.

TARGET MARKET CLIENTS DRIVE PORTFOLIO INVESTMENT DECISIONS.

The needs of the market, and those of the specific target market customers, drive organizational change. In response, executive management defines their corporate strategies to redirect organizational resources to give their companies a competitive advantage in their markets.

However, the reality is that there are always more things to do and spend money on than the organization has the resources to support. In short, the executives define a "portfolio" of potential business change options they must choose from when making their investment decisions. Since business change impacts business enablers, a mature organization makes investment decisions on the mixture of change recommendations that optimize achievement against their strategic goals and objectives.

ORGANIZATIONAL CHANGE IS PROJECT-BASED.

Making a change to business enablers involves a unique scope of work, has cross-organizational and cross-functional impacts, and is initiated as a temporary effort that must start and end within a designated time frame. Therefore, the work involved in facilitating business change is project-oriented and managed accordingly. A mature organization defines programs to oversee the change initiatives and then divides the work among authorized project teams with specific charters that document and approve the scope of work and other constraints.

The organization starts with a market- or customer-driven strategy, but the change required is not limited to simply changing the solutions delivered. Instead, the organization must sift through a plethora of potential change requirements and options spanning the supporting business enablers. Ultimately, the organization must establish programs and projects to define, execute, and control the required organizational changes across all affected business enablers.

CUSTOMER AND BUSINESS NEEDS ASSESSMENTS ARE KEY TO SCOPING CHANGE.
As part of portfolio analysis, every business, government, and nonprofit organization must determine what products and services their customers want, and why. Customer needs assessments help the organization uncover the products and services their target market/clients want, the capabilities and functionality required, the prices the customers can afford to pay, and the benefits the customers expect to receive. Additional analysis is required to determine the potential ROIs and the competition's offerings.

As this information becomes available, the organization must perform a business needs assessment to determine how they can deliver the desired solutions to their customers. It is this internal business needs assessment that leads to the discovery of change requirements across the enablers of the business.

BUSINESS ENABLERS HAVE INTEGRATED DEVELOPMENT LIFE CYCLES.
Products and services under development all require mature development life cycle processes to ensure that the requirements are understood, and that the right deliverables are created in support of those requirements. That is what this book on SDLCs is all about. However, the same concepts are true for providing the enabling organizational assets noted in the ECM; they all have life cycles involving the same categories of work defined in the traditional SDLC model. However, in the enterprise change model, which is multidimensional, we also have to be concerned about how the business enablers interconnect across other business enablers. In other words, a change in one business enabler likely drives changes across one or more of the other business enablers.

For example, new technology may drive modification to operational equipment that in turn may drive software changes to systems that support it. In turn, those changes create new organizational needs, such as the development and delivery of new resources, skills, and training aids. Also, production and inventory processes may have to change to accommodate the new technology.

PRODUCTS, SERVICES, AND BUSINESSES EVOLVE; SOMETIMES RADICALLY, MORE OFTEN INCREMENTALLY.
Few things in business are truly static. Change is inevitable. Therefore, it's important that the organization create a culture that thrives on change (people). The people in such a culture use knowledge and science for competitive advantage (technology), seek maturity and continuous improvement (process), understand that knowledge is power (information), and also understand their ultimate purpose is to fulfill market needs (customers and solutions).

In actual practice, this implies that the life cycle model as a whole, for each business enabler, is an iterative process. In some markets, particularly capital-intensive environments, this evolution might occur on an annual or even long-term basis. In other industries, such as online retail and information services, customer-driven product change occurs over hours or even minutes.

That's not to say that radical change, traditionally associated with business reengineering concepts, is not possible. However, radical change is extremely disruptive to the organization and typically motivated by a need to completely transform the business. Drivers for business transformations include the need to support new markets to sustain the business or implement radical process improvements to achieve competitive advantage and better support their customers. That's not to say that a transformation initiative always happens overnight. Some business transformation projects play out over several years in fulfillment of a new corporate vision.

Life Cycle Alignment

The depiction of the business enablers having identical life cycle phases is not meant to suggest that the phases occur in parallel and lockstep. Instead, the main intent of this visual model is to show that each of these enablers has a similar though independent life cycle. Moreover, it is important to note that changes in one enabler likely have an impact on one or more of the other business enablers.

Dependencies between the business enablers (see Figure 25) cause a ripple effect in change across the enablers, and not necessarily in synch with the same life cycle phases. For example, an approved change in a business process might drive future changes in the information systems used to support the modified process. When completed, the new or modified systems help reduce process errors and improve efficiencies.

The opposite can be true as well. For example, suppose end users work with an Agile development team and discover that the new application can evolve in a way that makes their jobs easier to do. In this case, changes in the underlying information system can drive changes to their affected business processes.

When business processes and supporting systems change, the organization needs to evaluate the roles, responsibilities, and skills of the people using the new process to ensure that they can use the system properly and execute new or revised business processes effectively. If the organization lacks the requisite skills and resources, they may have to hire new staff and develop and deploy new training aids. In some cases, outsourcing work may be a viable and cost-effective option to support a new change requirement.

While the requirements and design for the new processes and supporting systems may run in parallel, it may not be possible to define the new skills and design the related training courses to support new systems and processes until after those items are in the development or even testing stages. Prototyping and user experience design help mitigate the impacts on people and processes during the evolution of a system. Regardless, the design phase for training and support planning somewhat lags the design and development of the evolving processes and solutions, which can affect the timing of releases.

Performance testing is another area that can affect the release of a product. If the organization cannot sufficiently mimic the production environment, then it runs the risk of deploying a product with defects, and the company may have to recall products to fix them. Similarly, if the test environment cannot mimic the volume and loads expected in the production environments, the system may be deployed and fail if it cannot keep up with the operational demands. So, time must be put into the schedule to conduct proper production testing and break-fix activities after user acceptance testing.

Dependencies Drive the Schedule of Activities

As noted in the previous subsection, the life cycles for each of the business enablers will not likely all line up as depicted in the ECM graphic but instead link up through their dependencies. So, for example, new process requirements defined in the requirements phase under the Process enabler, drive functional and design requirements in the solution/system category of business enablers. In this case, the solution is a business system that supports the modified process.

The same is true for the Information Enabler requirements, which fall out of the process analysis activities. It is important to note that Process and Information enabler requirements analysis often go hand-in-hand, as changes in either can impact the other.

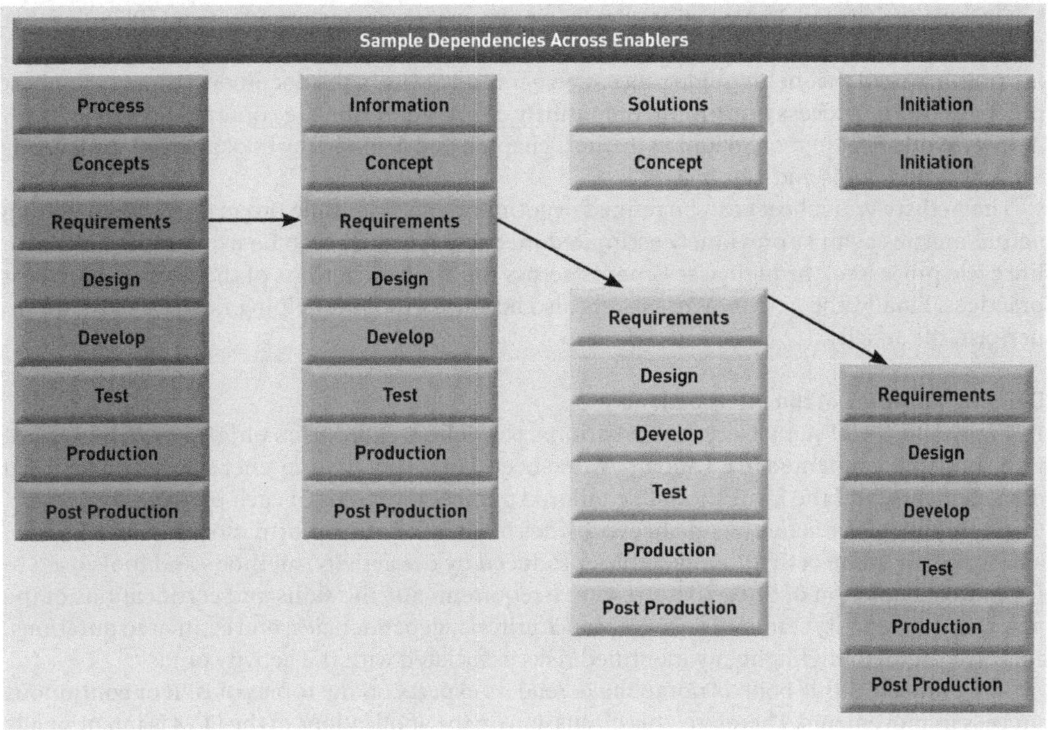

Figure 24. Dependencies across the enablers.

On the other hand, technical requirements cannot be fully established until the business and user requirements are defined and analyzed. The proposed process, information, and related technology changes also drive changes in the People enabler, by driving the need for new skills and related training to be developed. Here again, the training artifacts cannot be designed and developed until the business and technical requirements, design, and development activities have been completed. Figure 24 provides a graphical view of the potential impact of dependencies across the enablers.

Project Planning and Collaboration

Ideally, as part of the organization's process maturity program, assigned program and project teams should have a set of reusable processes and activities defined to perform the work across each affected enabler. If not available, the program and project teams have to define them, which adds to the work and the risks.

Most importantly, the program and project teams need to collaborate with other stakeholders, who have expertise in potentially affected business enablers, to understand their dependencies. As the activities and dependencies are defined, project managers assign resources to the planned tasks and work with the project team members and other stakeholders to estimate task durations. Collectively, the integrated life cycle activities, dependencies, durations, and available resources become inputs to the developing schedule plan.

Such collaboration during project planning is standard practice for project managers. What's novel in an enterprise change initiative is the notion that the work is not limited to developing an IT-based solution; rather, the organization needs to look across all organization assets that help enable the business to fully understand the total scope of work required to support a new strategic change initiative.

Enterprise Change Maturity

If the organization's ECM framework and related ECM processes are mature and maintained, each life cycle phase is supported by a set of processes to support the execution of new business strategies. However, if the processes are not well defined, the organization should start by capturing the best practices they have in-house as part of their quality and process maturity programs.

One way to do this is to have your process owners, process improvement team (PIT), or other SMEs document their existing best practices using an Activity Worksheet. The Activity Worksheet is a powerful tool that helps guide process owners and SMEs in the documentation of their best practices on any process improvement maturity effort. I describe the content and format of an Activity Worksheet in more detail in the next chapter, and again in my books titled Frameworks for Agility and Tools and Templates.

The Activity Worksheets can be refined over time, across multiple projects, by documenting actual metrics against previously estimated metrics. They can also be used to help improve life cycle processes, through assessments across multiple executions of the documented best practices. Finally, the Activity Worksheets also become a tool for helping novices learn how to perform the activity.

Defining Activities at the Intersections

In Figure 25 - ECM Activity Decompositions, the yellow call-out box highlights the type of information documented in an Activity Worksheet. This is not a comprehensive list of potential data elements, and the form should be tailored to meet the needs of each project.

A comprehensive activity worksheet provides fields to capture information on activity name, a description of the activity, deliverables produced by the activity, methods and tools used to facilitate completion of the activity, resource requirements, the skills and certifications of the practitioners, quality standards or acceptance criteria, dependencies, and estimated durations. It is also useful to highlight any identified risks associated with the activity or task.

The purpose of this book is not to make readers experts on the topics of BPR or continuous process improvements. Therefore, this discussion on the applications of the ECM is intentionally somewhat limited. However, if you have one takeaway from this chapter, please let it be an understanding that change in support of evolving business needs is so much more complex than simply creating new products or services. The entire organization acts as a complex system, requiring an as-is, to-be, and gap analysis across all affected assets that enable the business. Such analysis is the topic of the next subsection.

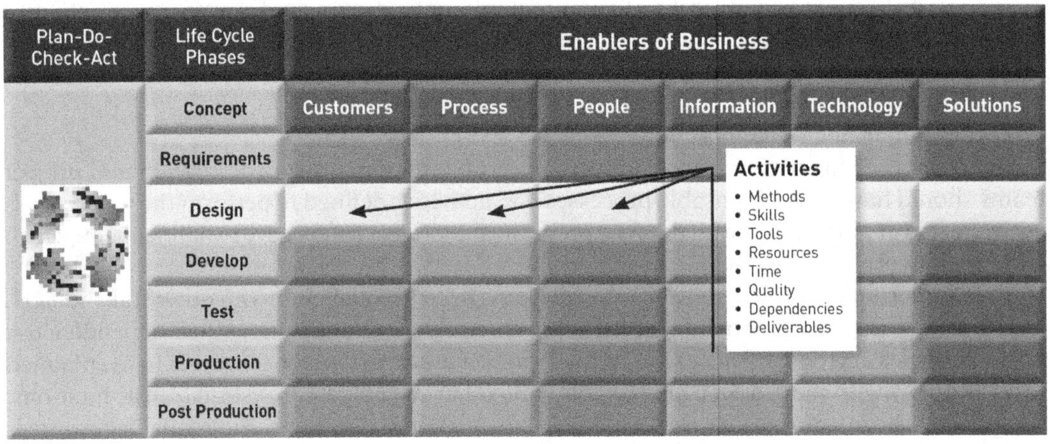

Figure 25. ECM Activity Decompositions

Change Disrupts the Life Cycles of the Business Enablers

When an organization contemplates a major change initiative, the potential impact is not limited to changes in the products and services offered by the organization. We have noted throughout this chapter that many if not all of the business enablers listed in the ECM are likewise affected, disrupting their life cycles. Therefore, each of the business enablers must be analyzed from a life cycle perspective to determine how much change is required.

Management consultants and business analysts often conduct a *needs assessment* to compare current capabilities and performance with potential or desired capabilities and performance across each business enabler. If the initial state of a business enabler is not well known, an *as-is assessment* documents existing capabilities and performance metrics. Next, a *to-be assessment* determines the requirements for the proposed future state, documenting desired capabilities, acceptance criteria, and performance metrics. Acceptance criteria are the elements, standards, and descriptions of the final deliverables or outputs of the activity.

The next stage is to conduct a *gap analysis*, which is interchangeable with the term Needs Analysis. The gap analysis is a study of the delta between the current and futures states to determine the scope of change required, alternative approaches to achieve the change, the impact of each alternative in costs, schedule, resources, and risks, plus recommendations on how to proceed.

A *CONOPS* investigation, during the concept phase of a life cycle, is a viable tool and approach for conducting, documenting, and reporting needs assessments and analysis. Chapter 12, Concept Phase, includes a sample CONOPS template. A CONOPS investigation is mission-oriented and stated from the view of end users.

Business Processes Interconnect Business Enablers

Organizations cannot look at the business enablers as individual silos. Everything has to work in concert to achieve the mission and strategic goals of the business. It is the function of business processes to integrate work and manage the flow of information across the enablers of the business.

For example, sales and marketing activities drive customer orders. Customer orders may be documented via a commercial website, or via an order entry system. The orders drive inventories of materials, drawing down from inventories and generating new orders for suppliers. The orders likewise drive production and delivery processes. Business systems are created to enable some or all of these cross-cutting business processes.

Business processes span the enablers of the business to coordinate the cross-functional and cross-organizational activities necessary to run and control the business to achieve corporate goals and objectives. Processes can be efficient, inefficient, or even broken. Broken and inefficient processes must be fixed before attempting automation via information systems.

Business systems can make business processes more efficient and the execution of activities more timely, exact, and coordinated, but only if the process is not broken in the first place. If the business processes are inefficient or wrong to begin with, automated business systems can make things even worse. For example, automated systems execute incorrect processes faithfully, and if left unattended may not be immediately detected, causing additional costs, delays in fulfillment, and other operational inefficiencies or failures.

Therefore, any business change initiative must look at the impact of cross-functional and cross-departmental business processes. Changes to the business enablers can affect the business processes, and vice versa. Moreover, systems that support modified business processes have to be updated.

CHAPTER SUMMARY

In recap, strategic or enterprise change initiatives potentially impact more than just the solutions offered by a company, government agency, or nonprofit. We also need to evaluate the impact on organization assets that enable the business to design, create, and deliver those solutions. Just as with software and systems, business enablers have life cycles, and we cannot shortchange the overall development or change process without putting the organization at risk.

The ECM is depicted in a matrix form as a multidimensional change model to demonstrate that many types of organizational assets are potentially impacted by a change initiative. The organization's assets are too numerous to be shown easily in a graphical model, so for simplicity, we categorized the business enablers by Customers, People, Process, Information, Technology, and Solutions.

We also learned that there are dependencies between the business enablers that often drive change across the enablers. Nevertheless, the life cycle for each enabler is independent and they may not all occur in lockstep fashion. An organization is wise to conduct a needs assessment study to evaluate as-is conditions, the desired to-be state, and perform a gap analysis to determine the scope of change required across all the business enablers.

The work to manage an enterprise change initiative has the same characteristics as those of a project. Both endeavors have a relatively unique scope of work, a specific beginning and an end, and an evolving understanding of the risks and scope of work throughout the engagement. Therefore, a PMO should oversee enterprise change initiatives as a portfolio of deliverables-based projects spanning the business enablers affected by the change.

Finally, the organization should include life cycle change processes in their process improvement and maturity programs. Activity Worksheets provide a useful tool for such purposes.

The next chapter provides an introduction into the common elements found within a standard or corporate SDLC.

11. Basic SDLC Phase Information and Artifacts

For those readers who want to know how taxonomies are used to classify and organize SDLC processes and the types of information commonly associated with IT processes and activities.

CHAPTER OBJECTIVES

At the end of this chapter, the reader should be able to discuss the following:

- Explain how work and deliverables are different but related concepts
- Provide a list of sample IT processes
- Describe a basic taxonomy for an SDLC phase
- List the primary information elements contained within SDLC process descriptions
- Discuss how to write phase descriptions, goals, and objectives
- Explain the difference between roles, responsibilities, and skills
- List and describe the terms that make up the RACI and RASCI acronyms
- Define the terms process, activity, tasks, and work package
- Explain the purpose of SDLC phase or milestone reviews
- Explain the purpose of entry and exit criteria
- Explain the purpose of phase review checklists

AUTHOR'S INTRODUCTION

This chapter describes a typical SDLC phase-based structure, including the processes and artifacts that provide instructions and information necessary to execute the type of work associated with the phase. Again, it is important to think of an SDLC phase as the means to classify work, not as indicating a sequence of events. The inputs and outputs or dependencies of processes, activities, and tasks generate the sequence of work.

A mature SDLC can have hundreds of pages of descriptive content. I worked with one government agency that built a standard SDLC with more than 1700 pages. The agency is very large, with several divisions and hundreds of locations and operating groups. They also must deal with government mandates that explicitly define the government services they offer.

The effort to build a standard SDLC, document other corporate processes, and enforce IT governance policies agency-wide, came about as the agency found it increasingly difficult and expensive to support the many technologies and redundant software applications and systems implemented over the years. Though a federal agency's ultimate executive oversight group is the Congress, the agency feels no less pain when they fail to deliver a software application or system on time, on budget and in conformance with the mandates.

The agency has since moved forward to implement Agile practices, and to reduce much of the documentation and other administrative burdens. Still, they maintain an extensive library of standard life cycle processes for systems development, project management, human resources, acquisitions and supplier management, and program planning, budgeting, and execution.

What's important to understand is that the implementation of Agile practices does not eliminate the need to develop and maintain standardized processes. Conceptually, Agile practices applied to IT encompass only a small part of an organization's life cycle process requirements, and typically do not address good engineering practices at all. Moreover, you may find it difficult to master the organization's standard life cycle processes if you don't understand what to expect, and how to read and use the processes.

In previous chapters, we noted that SDLCs are typically implemented uniquely by organizations to fit their business needs, and for competitive advantage. As a result, this chapter cannot describe every SDLC process, activity, and the related artifacts that readers might encounter. Fortunately, basic SDLC processes and associated artifacts and information have more features in common than not, across many if not most SDLC formats. This chapter introduces common terms, tools, and SDLC concepts.

BASIC DEFINITIONS FOR DESCRIBING WORK AND DELIVERABLES

Throughout this book, I've used the terms *phase*, *activities*, and *tasks* to describe different levels of work. It's not unusual to hear these terms used interchangeably, as if they mean the same thing; but as you may recall from Chapter One, they do not. Activities provide summary level details of underlaying tasks, usually for reporting purposes, while tasks are descriptions of work.

There are other terms used commonly in IT product development, and in the context of managing IT-based project management life cycles, which have specific semantic meanings. Let's deal with many of these terms and definitions now, so that we have a common semantic meaning for each going forward.

Artifacts: In a modern and IT context, *artifact* is a term that describes a collection of things that are made and reused across a project or an SDLC. The word derives from two Latin words: *arte,* which means "with skill," and *factum*, which means "to make something."

An artifact can be a template, document, plan, diagram, chart, software code, test scripts and test results; descriptions of processes, methods, activities, and tasks; or virtually anything that is created by or in support of a project.

Phase: This term has two meanings in the context of an SDLC. First, a phase can connote a specific set of processes and tasks associated with a type of life cycle-oriented work, such as the design or development phases in the traditional waterfall SDLC model.

The term phase can also imply a scope of work that is related and collectively completes a major deliverable of the project. For example, a complex system may have a phase of work devoted to developing hardware components, while another phase of work is devoted to development of software or firmware related to the hardware.

Activity: A higher-level collection of work (e.g., a set of tasks) that creates a specific deliverable (product, service, or result). An activity is not a deliverable; rather, it is a collection of tasks required to produce a deliverable item. The purpose of defining work as the activity level is to organize and manage work used to create a specific deliverable item in a project, and to aggregate resource and costs information associated with the deliverable for management reporting purposes.

For example, we might have an activity specified to collectively manage and report on the work associated with developing a BRD within the requirements phase of a project. However, we can also define an activity to manage deliverables-oriented work and information at a lower level of granularity of that same activity. In such a case, we might have a lower level activity to schedule, control and report on the tasks necessary to run JAD sessions to capture business and user needs for the business requirements specification. So, in effect, activities can be nested within a WBS decomposition.

Task: Discrete pieces of work that roll up into higher-level activities. Each task must have a deliverable outcome that is the focus of the work, and that is quantifiable and measurable to indicate successful completion. Tasks have dependencies to integrate and define the flow of work across subtasks. Task descriptions provide general information on procedural methods, tools, resources, skills, task duration, and risks.

Work Packages: Used in the project management community to refer to the lowest levels within a WBS that produce a deliverable but are impracticable to further decompose. As with an activity, resources and costs are rolled-up for reporting purposes at the work package level.

The work package concept reduces reporting burdens by hiding the details of the underlying tasks. On the other hand, from an SDLC perspective, the problem with using the work package concept is that it hides the details of the underlying work. To overcome this problem, the work package definitions must point to the relevant SDLC and other corporate processes that define the organization's best or approved practices and standards.

Another issue I personally have with the use of work packages is that work is performed at all levels of a WBS, and not just on the lowest definable deliverables. I discuss this issue in greater detail in Chapter 12.

Deliverable: Tangible or intangible things that are defined, created, or modified by the work performed during a phase, activity, task, or work package within a project. Hierarchical decompositions of the project's highest-level deliverables define the total scope of a project. A deliverables-oriented hierarchical decomposition is referred to as a WBS.

As an example, a project's charter specifies the primary deliverables that must be produced by the project team. The project team then breaks apart the primary deliverable into smaller component deliverables to better understand, define, and manage the overall work. Other deliverables may be required to support organizational management and oversight policies.

Process: Typically defined as an ordered progression of activities taken together to produce a specific result. In the context of systems development and project management, a process is a series of steps taken to develop the deliverables of the project (e.g., product, service, or expected outcomes) plus the activities and tasks that are necessary to manage the project effectively.

However, a process is not a schedule. A process description is generic, reusable in the application, and typically conveys a set of best practices describing how to perform the work. So, while a process may specify a preferred ordered sequence of work tasks or procedures, it does not contain a specific start or end date, resource availability information, or other constraints that are necessary to define its use on a particular application or project schedule.

Schedule: A tool used by project managers and team members to define the flow of work over time within the unique scope, time, resource, quality, and dependency constraints of a project. A schedule should incorporate best practice processes, whenever they are available. To plan and convey the actual flow of work over time, we need to show the dependencies between various deliverables, phases, activities, and tasks and include the constraints of time and resources.

How do we then explain the difference between a schedule and a process? That's simple. A process is an ordered set of activities and tasks, describing a repeatable practice, whereas a schedule is a uniquely applied set of activities and tasks executed in accordance with established dependencies, durations, and resource availability.

As an example, business requirements may be defined as a deliverable within a project's charter and WBS. The project's business analyst leverages the documented process, tasks, and related best practices defined within the organization's SDLC process for requirements gathering to perform the work correctly and efficiently. The business requirements as a deliverable are uniquely applied within the project schedule by noting the start and finish dates or durations, resources, and dependencies with other processes.

WBS: The classic and most common definition of a WBS is a deliverable-oriented, hierarchical decomposition of a project's work. The point of a WBS is to decompose the highest-level deliverables into smaller parts to expose the total scope of work required to complete the project. The decomposition process starts with breaking the highest-level deliverable into its component parts, and then eventually exposing the work as related activities and tasks at the work package level.

Defining activities and tasks within a WBS can get tricky. For example, what happens when a development team must describe the work associated with assembling several parts at one level of the decomposition when those subassemblies also include components that require assembly or fabrication? My recommendation is to create a separate WBS for each subassembly, and then integrate them all in activity network diagrams and Gannt charts.

As a real-world example, if we were creating a WBS to describe the work associated with assembling an aircraft, we would logically start with the highest-level deliverable being the aircraft itself. Next, we would list as deliverables-oriented decompositions under the aircraft major assembly components, such as the fuselage, wings, tail assembly, wheel assemblies, engines, electronics, hydraulics, and interior elements.

Each of these subassemblies logically decomposes into other subcomponents and work. For example, an aircraft wing includes:

- Frame
- Skin
- Fuel tanks
- Hydraulic lines and subsystems
- Electrical wiring harnesses
- Fuel tanks, fuel lines, and subsystems
- Auxiliary power unit to supply electrical power and air conditioning to the plane
- Control surfaces such as ailerons, flaps, speed brakes, and spoilers

Unless purchased from another vendor, each of these wing subassemblies and systems might have separate WBSs to describe the deliverables and work associated to fabricate or assemble them. This process of decomposition and separation of work continues until we get to the lowest practical level to define.

At that point, if not concurrently, we begin to define the activities and tasks for creating the subassemblies and then installing each of those subassemblies into the higher-level components that eventually make up the aircraft as a whole. In other words, while a WBS exposes the scope of work for the project as a set of deliverables, it is the activities and tasks that define the methods, tools, resources, dependencies, and time required for each deliverable item.

It's important to note that a WBS is not a process, as it merely shows a deliverables-based decomposition of the things created by the project team. Also, a WBS is not a workflow, as it does not show the actual sequence of events over time. Likewise, a WBS does not show repeated sequences of activities and tasks, such as occurs during an IID sequence repeated during each release cycle.

For each WBS, the project team must define the activities and tasks necessary to produce the deliverable items. The defined activities and tasks will have dependencies, and those dependencies establish the flow of work, plus they describe how to conduct the work necessary to create the deliverables.

WBS Dictionary: This term is used in the project management community to describe a document or worksheet that contains information on each WBS component. In this book, the Activity Worksheet provides this capability in a structured and reusable format. In other words, use the Activity Worksheets to create your WBS Dictionaries.

IT ORGANIZATIONAL PROCESSES

The organizational processes developed to support a large IT organization can be quite extensive and comprehensive and will span more than just SDLCs. For example, IT processes can span acquisitions, change management, communications, enterprise architecture, financial controls, human resources, IT governance and strategy, maintenance, operations and infrastructure, portfolio management, program and project management, risk management, supplier management, support services, and, of course, systems development.

Smaller companies typically have limited resources to build out such comprehensive IT process documentation. However, that does not mean the organization cannot begin to develop some basic process descriptions and documentation, and then build out the subprocess and activity level documentation over time. The Activity Worksheet introduced in the previous chapter is a great tool for this purpose. The information elements contained within an activity worksheet are described in more detail later in this chapter. Table 15. Activity Worksheet Information.

Information Classifications

Whenever an organization needs to document business processes and related information systems, where common practices need to be documented and made available in a distributed manner, it is useful to define a classification scheme upon which information of a similar type is maintained and managed throughout the execution of the process. A taxonomic approach to information classification and organization is common in many SDLC frameworks.

Taxonomies

The word "taxonomy" came from the French word taxonomie, which is derived from the Greek word taxis, meaning arrangement or division, and nomos, meaning distribution, arrangement, or management.[111] Most of us have learned about the subject of taxonomy from our high school biology classes. The practice was originally developed by Swedish scientist Carl Linnaeus, as a standardized naming system for animal and plant species.

However, in a more generalized sense, taxonomy refers to the science or technique of classification. In other words, the science of taxonomy is much broader than zoology, providing methods for classifying knowledge as collections of information about virtually any complex subject or system.

Taxonomies typically take the form of hierarchical structures for managing information related to the subject or system. A taxonomic decomposition breaks the highest-level subject or system into lower-level components, with increasingly detailed information maintained at each level of the decomposition. In other words, each layer of the decomposition exposes greater details about the subject as a whole, including information on the various components or subsystems that collectively belong to the subject.

Figure 26 provides a graphical view of a hierarchical decomposition for an SDLC phase. Once formed, a taxonomy provides a conceptual framework to manage, analyze and access information related to the SDLC processes.

From a graphical perspective, an SDLC phase looks a lot like a WBS. However, they are different in composition and have different purposes. A WBS is a decomposition of the deliverables associated with a product under development, exposing the work associated with developing a specific product within the hierarchy. In contrast, an SDLC is a generic decomposition of product or systems development work, in the form of standardized processes, activities, and tasks associated with developing the product.

Note, it is a mistake to equate or combine the taxonomic structures of a WBS with those from the SDLC. Doing so enforces the waterfall-based concepts that development phases operate sequentially. The WBS simply indicates the scope of product development work from a deliverable's perspective, while the SDLC provides a method to classify, organize, manage, deploy, and use standard life cycle development processes and associated activities and tasks. In other words, an SDLC process provides detailed information on the work and deliverables associated with the execution of the process.

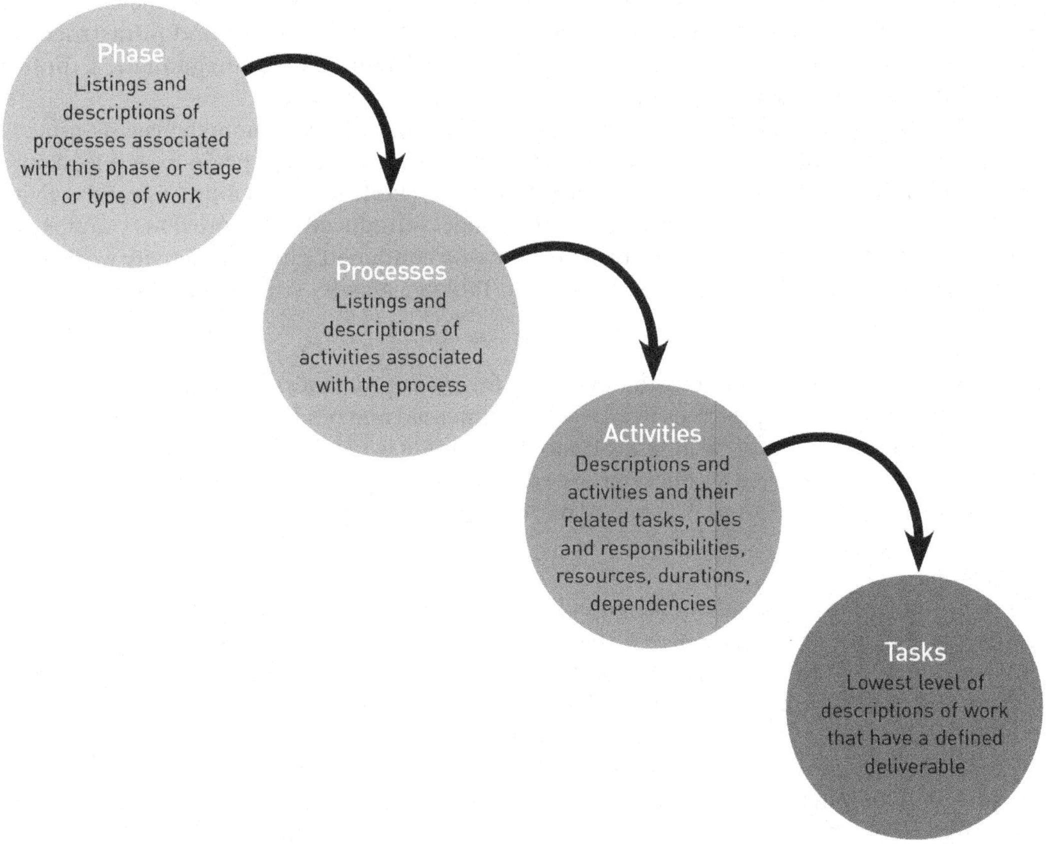

Figure 26. Taxonomic View of an SDLC Phase

AN SDLC TAXONOMY

The term artifacts refer to the types of tools and documents managed within an SDLC framework, consistent with the SDLC taxonomy/classification schema installed by the IT organization. Loosely speaking, artifacts are human-made tools or other objects that require some level of specialized skills to make or use.

In our SDLC taxonomy, the phases represent the highest level of classification. Some people refer to the higher-level SDLC process classifications as stages. Each SDLC process includes the listed information artifacts, as shown in Table 14, which provide further details governing the proper execution of the associated processes.

This classification pattern, or one very similar to it, repeats across all the phases included in a standard SDLC framework.

Table 14. SDLC Phase Artifacts

SDLC Phase Artifact	Descriptions
SDLC Phase or Stage	Any of the SDLC phases; each includes a set of processes that are required to complete the deliverables associated with the type of work classified therein.
SDLC Process	A set of activities and tasks performed in a specific way to accomplish some part of the work associated with the phase and, in a mature organization, also describes the entity's set of best or "good" business and engineering practices.
Activity Worksheets	Used during a process maturity program to document SDLC and other corporate processes, activities, and tasks. Once completed, they provide specific guidance to novice practitioners on each listed activity or task.
Activities	Activities represent a higher level description of work where information is collected for management reporting purposes. As with tasks, activities must have one or more deliverables associated with the completed work, and with measurable acceptance criteria to define expected levels of quality. Tasks are lower level descriptions of work where a deliverable can be defined. Flow Charts are typically in the form of an Activity Network flow diagram depicting the flow of work to accomplish a related set of work associated with the described business process. Descriptions detail the high-level outcome(s) of the stated process, activity, or task, plus objectives stated as progressive steps with performance metrics indicating successful completion of each. Goals and objectives list the final deliverables and related work to be accomplished throughout executing this process Responsibility Assignment Matrix (RAM) - a listing lists and describes the individuals involved in the process, and their roles and responsibilities. A RAM helps clarify the roles and responsibilities in cross-functional and cross-organizational teams; and, are often defined within the RACI or RASCI chart formats—Responsible, Accountable, Supporting, Consulted and Informed—while executing the described process.

COMMONLY USED SDLC PHASES

In this book, I have defined the following SDLC phases or stages as the high-level classifications to manage SDLC Processes:

- Concept
- Requirements
- Design
- Development
- Testing

- Production turnover/deployment
- Post Deployment

Each SDLC phase or stage represents a placeholder to classify and manage related SDLC processes by type of work performed.

SDLC PROCESSES

The SDLC framework provides the classification schema to manage SDLC processes in context with the defined phases and stages. While the names for SDLC phases or stages vary from organization to organization, the type of work associated with each is consistent. SDLC processes are mapped to the standard SDLC Phases, enabling the retention of related process information within the SDLC framework. The following list indicates the types of SDLC processes associated with each type of work.

- Concept—Processes to justify, analyze, validate, approve, and initiate organizational investments in developing new products
- Requirements—Processes to gather and document business, customer, and end user requirements
- Design—Processes to define and document the system architecture and solution design requirements
- Development—Processes to create the product's deliverables
- Testing—Processes to test the product deliverables to ensure conformance with the requirements, architecture, and design
- Production turnover/deployment—Processes to make sure the product can be supported and maintained in operations
- Post deployment—Processes to support ongoing bug fixes and enhancements, plus the eventual end-of-life processes

Activity Worksheets

An Activity Worksheet provides a common method and tool to capture, maintain and deploy information on SDLC processes, activities and tasks. This tool supports the implementation of a process improvement and maturity program. The worksheets also provide a guide for non- expert practitioners in the execution of the organization's defined processes, activities and tasks. I've also used the completed worksheets to help plan and build the resource and cost estimates for a project.

The goal is not to make documentation in the activity worksheets an onerous and time-consuming task, as the domain experts and practitioners will not support it. It is best to use this tool on-demand whenever a task is not well understood or documented. In other words, if a task is assigned but not well understood, have the assigned manager take a few moments to getwith a SME and document how they would do the work. The activity worksheets are updated over time, when and as the initial details and estimates prove insufficient or incorrect.

Tasks versus Work Packages

If you have spent any time working on IT projects, you may have wondered what the differences are between the term's tasks and work packages. Unfortunately, it is not unusual to see the terms used interchangeably.

DID YOU KNOW?

Useful Definitions for Work:

Task–a generic term for discrete units of work that have a definable and measurable deliverable.

Work Package—The lowest level of work identified in a deliverables-oriented WBS using the 8/80 rule. E.g., tasks should have a limited duration of between 8 hours and 80 hours.

Activity—A set of tasks or work packages used to roll up and report costs and resources around a meaningful deliverable item that is relevant to the customer.

From a definitional perspective, the word *task* implies a specific piece of work and a specific result. In the project management community, the term *work package* is reserved for the lowest level of work defined within a WBS. A work package can contain multiple deliverables and components of work. Therefore, a work package is better thought of as a "mini project" that defines the lowest level of work that is practicable to define, schedule, and manage on a project.

DID YOU KNOW?

Microsoft, in their scheduling application Microsoft® Project®, uses the term task to define work, and the term summary task to roll up collections of tasks.

It is left to the project scheduler or project manager to determine why certain tasks are rolled- up into a summary task.

A WBS is a deliverables-oriented and hierarchical decomposition of a project into smaller and more manageable components of work, with each level of the WBS having a defined deliverable. Each level of decomposition represents work associated with creating a unique deliverable, but the lowest level work packages represent the smallest level of work that can be scheduled and budgeted.

Project managers use the "8/80 rule" to define work packages as a scope of work spanning anywhere from 8 to 80 hours duration. (e.g., between 1 to 10 workdays). Each work package is self-contained, has a budget estimate, and is managed as a separate task on the project's schedule.

The terms tasks and work packages may be synonymous and interchangeable. However, there is no reason why tasks defined within an SDLC process must be limited to the 8/ 80 rule. In fact, that strategy may confuse your customers if it's not linked to a specific deliverable that your customers or executives can recognize. Also, within an SDLC process description, it's more important that the underlying tasks spell out how the work within a process is to be accomplished. Organizing work as a decomposition of higher-level deliverables does make it easier to visualize the composition of the final deliverable. And, for that reason, it is often practical to use the WBS elements as the basis for activity reporting.

One final comment on this topic. A WBS does not provide any practical information on the actual scope or flow of work. A WBS looks more like an exploded parts list than a process flow diagram, network diagram, or Gannt Chart.

Project Reporting

There are important reasons to organize work as collections of tasks or work packages within a project schedule. Those reasons have to do with budgeting and reporting requirements around major deliverable items.

In large projects with hundreds if not thousands of individual tasks, it is rather impractical to track and report budgets and work completions at a task level. We need a method to roll up the resources and costs associated with related sets of tasks. However, don't be surprised if your customers roll their eyes at you if you try to explain the relevance of cost reporting on 8 hour to 80-hour work increments. They are generally much more interested in the costs and resources consumed to create a specific and measurable deliverable that has meaning to them, and not work reported based on some arbitrary rule.

Therefore, throughout my career, I've made a firm distinction between tasks, work packages and activities to indicate discrete pieces of work specifically versus useful aggregations of work for reporting reasons. In other words, the word task implies a discrete set of work with an identifiable deliverable, work packages break out work into smaller increments for project scheduling purposes, and the word activity is used to differentiate an organized collection of related work for budgeting and reporting purposes that has a meaningful deliverable item to the customer. An activity can be a work package, but an activity can also be an aggregation of work packages that define a higher-level deliverable and important to the customer from a budgeting and reporting perspective.

Flowcharts

Flowcharts depict the flow of work within an SDLC. In contrast with hierarchical views of work, which graphically depict work that is somehow related, flowcharts show the sequence of work. The sequence of work derives from the dependencies among processes, activities, and tasks.

However, those dependent relationships are impossible to see in a hierarchical diagram, and difficult to visualize from a review of the dependencies. A flowchart addresses both of these visualization issues.

In project management, we create project schedules through a three-step process:

- **Determine the necessary processes, activities, and tasks, and their dependencies**
- **Use dependencies to define the flow of work within an activity network diagram (a.k.a. arrow diagram or a PERT Diagram)**
- **Input resources and durations to form the project schedule and usually displayed in a Gantt Chart view.**

Descriptions

Process, activity, and task descriptions provide a general overview of the work to be performed, and the expected results or deliverables. Within an SDLC, process descriptions are text-based and written on the activity worksheets or SDLC process (activity, task) description sections.

Goals and Objectives

Goals are general descriptions of the expected outcome of the process, activity or task. They provide clear-cut statements of the desired results, without the burden of explaining the methods used to achieve the desired outcomes. Well-written goals keep the focus on clearly stating the purpose of the work.

Objectives also represent desired results or outcomes, but at a more granular level. Objectives are stated as activities and sub-goals to help make clear the steps necessary to achieve the overall goal. Specifically, each objective articulates a single step in a progression to achieve the overarching goal.

But be careful, an objective is not a step. Like goals, an objective is expressed as a desired outcome, just at a more granular level of detail than the high-level goals. In contrast, steps are explicit activities or procedures. If there is the potential for confusion on how work is performed to accomplish the stated goals and objectives, then the Activity Worksheet should also include a section called Procedures, or Method, with a list of detailed steps required to achieve each stated objective.

In summary, goals establish target outcomes for SDLC processes, activities, and tasks, while objectives state the steps necessary to achieve the desired outcomes.

As an example, the goal of unit testing is to ensure that software code is tested incrementally, as written, before integrating it with the baseline code where it will be more difficult to find and fix any defects. The stated objectives of unit testing might include create test cases; create test data; determine expected return values; create tests scripts (to run the programmer's code through the test cases most efficiently); execute the tests; and, fix defects in code.

Responsibility Assignment Matrix (RAM)

A RAM provides both a method and a tool for assigning responsibilities of people who are necessary to complete a process, activity or task. There are many variations of this practice. In this discussion, we focus on the RACI and RASCI forms. Both terms are similar, and both are used extensively in the IT industry. The acronyms stand for the following roles:

R - Responsible: The person who is responsible for carrying out the task. The responsible person may also hold the accountable role, or as designated by the responsible agent.

A - Accountable (a.k.a. Approver): The person who is held responsible for ensuring completion of work within the allotted constraints of scope, time, cost, resources, and quality. Only one person holds the Accountable role.

S - Supporting: One or more persons who assist in the execution of the process, activity or task. Think of this category as a resource pool to help complete the work more quickly.

C - Consulted: A person or persons who have domain or subject matter expertise and who help inform the project team members or appraise their work during the execution of the process, activity, or task. They do not perform the work, however.

I - Informed: Individuals such as executive managers, customers and other stakeholders who need updates on the status of the process, activity, or task. They do not perform the work, though they may approve and received the deliverables, or they may be affected by the work or deliverables.

ACTIVITY (PROCESS, TASK) WORKSHEETS

The Activity Worksheet is an extremely valuable tool to help define and execute corporate processes and best practices. Completed as part of an SDLC Frameworks for Agility or improvement program, such as CMMI, once completed, the worksheets become part of the SDLC framework, within the phase(s) with which they are associated.

Activity worksheets provide descriptions of work. As such, they can be used to define work at the process, activity, or task level. If you want to call them a Process Worksheet or Task Worksheet, there's no problem with that. However, the general work information documented across processes, activities and tasks are similar, so the template applies to all three levels of work decompositions.

I use the term Activity Worksheet because that's the lowest level of reporting requirements on a project. But I always initially develop the activity worksheets at the task level and then use them to build up to the process level. In other words, the activity level worksheets provide a summary level description of underlying tasks, and the process level worksheets providing summary level details of the underlying activities and tasks.

Table 15 lists and describes the type of information gathered and maintained within an Activity Worksheet. The template provided has a standard format. However, users should not be afraid of modifying, adding, or subtracting information elements on this form to meet their unique needs.

Table 15. Activity Worksheet Information Components

Information	Descriptions
Level of Work	Specification of whether the work is a process, activity, or task.
Standards	List applicable industry and corporate standards that apply to this work.
Procedures or Methods and Tools	Documentation of the approach or steps necessary to complete this activity, plus any tools necessary to perform the work.
Roles & Responsibilities	A RACI or RASCI matrix listing participants and their roles and responsibilities.
Deliverables	Tangible or intangible work items, typically represented as a product, service or result.
Work Description	Instructions on how to perform the work.
Skills	Human capabilities developed from experience, certifications, education, and training.
Resources	Materials consumed during the execution of a process, activity or task, or that become part of the deliverables.
Time Durations	Estimate durations to set up, complete, and shut down the activity or task.
Setup	Tasks required to set up equipment and tools, and otherwise prepare to execute the process, activity or task.
Quality	Applicable standards/acceptance criteria and metrics that demonstrate conformance with requirements.
Dependencies	The other processes, activities, or task the work is linked to, and how. Dependencies are used within an activity network diagram, Gantt Chart, and schedules to define the flow of work across process and activities.

Take a moment to review and become familiar with the SDLC phase artifacts and activity worksheets as I'll refer to these items throughout my book titled *Frameworks for Agility*, which is devoted to describing in detail each of the SDLC processes across both traditional and Agile-based SDLC models.

Level of Work

This field in the activity worksheet indicates whether the information contained in the worksheet is a SDLC process, an activity, or a task.

Standards

In this section in the Activity Worksheet, list the industry and corporate standards that apply to the work described, such as:

- **IEEE Standards for Computer Technology and Software and Systems Engineering**
  - As of the date of this writing, IEEE maintains 197 Computer Technology Standards, 25 Healthcare IT Standards and 2 Software and Systems Engineering Standards.[112]
- **IS Audit and Assurance Standards (ISACA)/ Control Objectives for Information and Related Technologies (COBIT) or other IT Governance (ITG) standards**
  - International Organization for Standardization (ISO)/ International Electrotechnical Commission (IEC) 27001 standards for developing an information security management system (ISMS)
- **AXELOS Global Best Practices**
  - The AXELOS portfolio of best practices spans eleven methods, primarily encompassing IT service management, project, program and portfolio management and cyber resilience

- **Object Management Group (OMG) Standards**
  - As of the date of this writing, OMG has adopted and maintains 238 technology specifications, spanning 37 categories [113]
- **The Open Group Standards and Certifications**
  - As of the date of this writing, The Open Group has developed 167 open and vendor-neutral technology standards and certifications, plus they have developed over 2,000 publications that include many standards, though some are not listed in their current library [114]
- **Others**

Methods and Tools

Mature organizations work aggressively to ensure that they have standard practices, methods, and tools to improve SDLC performance and consistency across IT-based programs and projects. This section of the activity worksheet documents the approach or steps used to perform the activity. The section should also include a listing of items that help create the deliverables, such as computing hardware, networks, and software tools, instruments, equipment, machines, jigs, templates, and other necessary tools.

If corporate methods already exist and are documented, then the standard method is referenced in the worksheet. However, if there is not a standard and documented method, then the defined work needs to be fully detailed within the activity worksheet.

Roles and Responsibilities

This section within the activity worksheet provides a place to identify the skills and resources required to complete the work. There can be any number of people involved in performing, reviewing, and approving the work and acceptance of specified deliverables. Each participant may have different skills and experiences, depending on the type of work product delivered. A RACI or RASCI chart in this section identifies the roles and responsibilities of the individuals supporting the work.

Large program or project teams include team members from disparate functional groups and departments to provide all the knowledge domains and skills necessary to produce the deliverables associated with each identified process, activity, or task. The role of a person describes the position they have relative to the work conducted on the program and projects, whereas responsibilities define the work they are expected to perform or the authorities they have. Skills play an important part in the definition of responsibilities, as skills define the education, certifications, and experience the person should possess to successfully perform and complete the work assigned within an activity or task.

Deliverables

This section of the activity worksheet is where the outputs of the work are described. Deliverables must be definable and measurable to ensure acceptable results or outcomes. Sometimes the term "acceptance criteria" is used to define the standards or measures of quality that indicate successful completion of the work and deliverables.

Each life cycle phase, process, activity, and task produce a set of deliverables as a result of the work that is specified. Deliverables can take one of several forms, such as products, services, or outcomes. At the highest level of abstraction, these are the things that customers want. At the lower levels of a WBS, deliverables are the components that make up the higher-level deliverables. Some deliverables within a program or project apply to the management of the project, while others can be subcomponents of other deliverable work products. Deliverables can also include documents related to describing the requirements, design, system administration, policies, procedures or use of the final solution. In short, deliverables are the result of work.

Work Descriptions

In the activity worksheets, work descriptions define the scope of work that creates the project's deliverables (e.g., product, service, or outcomes). The descriptions should include goals, objectives, and metrics that define "what good looks likes" for each deliverable. In other words, we need this information to know if the work and deliverables are acceptable. Goals, objectives, and metrics collectively form the acceptance criteria for a designated scope of work.

For corporate business systems, the overall program or project deliverables, goals, and metrics should map back to the strategic objectives mandated by executive management. It is not enough to say that the program or project must accomplish great things; what's more important is that the investments in programs and projects align with and justify the corporate resources invested in them.

Skills

The processes, activities, and tasks associated with an SDLC are quite technical, and the people who execute them require specialized training, certifications, and skills. Therefore, the activity worksheet should call out the necessary training, skills, and certifications necessary to complete the specified activity.

Resources

In a generic sense, resources are any assets required to complete the documented process, activity, or task. However, we've already recognized human resource requirements in the Roles and Responsibilities section, and non-consumable resources within the Methods and Tools section. What remains are the materials that are consumed or become part of a product during the execution of a process, activity or task. For example, material resources can include electronic hardware components assembled within a final product, while chemical agents used to clean a product represent materials consumed during the execution of the work.

Time Durations

Activity and task durations are critical information elements when building a project schedule. In fact, along with the identification and allocation of shared resources, estimated times are necessary to turn a flowchart, such as an activity network diagram or PERT chart, into a schedule. In rare instances, time durations are deterministic, meaning that the outputs and behaviors during the execution of a process, activity or task are constant and predictable.

However, by definition, projects are unique and subject to uncertainty, which implies that the inputs and behaviors in a process are not well defined or understood. Therefore, the time durations are likely estimated and in fact may vary randomly within some probability distribution if executed frequently. In other words, the processes, activities, and tasks present stochastic behaviors. As stochastic behavior is the norm; a range of estimated times is enough when there is known variance in the estimates.

Setup

Some activities and tasks have extensive setup requirements that employ the use of equipment, tools, computing devices, and other resources, and take some amount of time to complete. When this is the case, the setup activities, time, and resources must be called out in the activity worksheet.

Documenting the setup process and procedures is important for two reasons. First, the work has to be accomplished to start the actual work associated with the defined process (activity, task). Second, setup tasks represent non-value-added work, as the effort does not directly contribute to the product's capabilities.

In other words, if we could somehow build the product without the setup time and resources, it would not change the overall value of the product, but it would reduce costs and improve throughput. Over time, mature organizations work to minimize and eliminate setup time, resources and related costs wherever practicable, as they represent waste and inefficiency.

Quality

When executing a process, activity, or task, we need to understand what characteristics demonstrate a successful result. Ideally, quality metrics should be quantifiable and measurable to be of value. However, qualitative descriptions are useful when quantitative measurements are not known or practicable.

Dependencies

Throughout this course, we have noted that dependencies drive the flow of work, and not the hierarchical relationships within the taxonomy of an SDLC Framework or a product's WBS. Dependencies can take the form of predecessor or successor events. As the terms imply, a predecessor event occurs before the activity or task, while a successor event occurs after the activity or task.

Beyond predecessor and successor relationships, four types of dependency relationships are possible.

DID YOU KNOW?

Process level dependencies are possible,

but dependencies are more commonly applied at the activity, task, and work package level.

- Finish-to-start
- Start-to-finish
- Start-to-start
- Finish-to-finish

Let's see how these relationships work. For all four scenarios, we have two activities, Activity A and Activity B. For all four situations; the predecessor is Activity A, while Activity B is the successor. The Finish-to-Start (FS) relationship (Figure 27) is by far the most common dependency relationship. The FS relationship is a strictly sequential dependency where Activity A must start and stop before Activity B can begin. Think of this as running a relay race where the first runner has to finish their sprint before handing off the baton to the next runner.

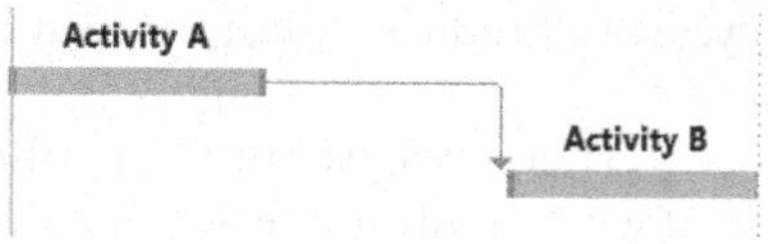

Figure 27. Finish-to-Start (FS) dependency relationship

In contrast, the Start-to-Finish (SF) relationship (Figure 28) is by far the least common dependency situation. In such a scenario, the predecessor Activity A cannot begin until the successor Activity B is complete. This may seem odd; but think of it from the perspective of a schedule plan or Gannt Chart, where the formation of precedent/successor relationships comes from their order of listing, and the Finish of the successor activity is dependent upon the Start of the predecessor activity. In other words, use this type of dependency when the successor activity or task can continue to proceed upon delays of the predecessor activity. For example, ticket scalpers—an afterthought if ever there was one—can continue selling their tickets when a baseball game is experiencing a rain delay.

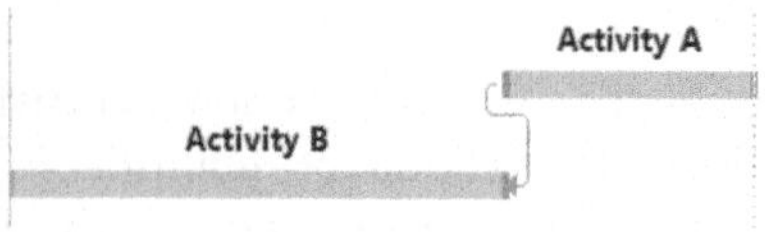

Figure 28. Start-to-Finish (SF) dependency relationship

Use the Start-to-Start (SS) relationship (Figure 29) when two activities can start at the same time and run in parallel. The SS relationship doesn't mean the activities also have to finish at the same time, only that they have to start at the same time. As an example, in a race, the two athletes start running at the same time.

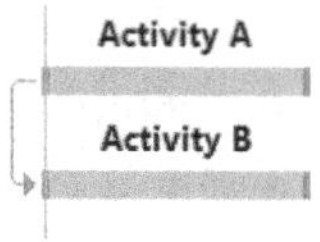

Figure 29. Start-to-Start (SS) dependency relationship

The Finish-to-Finish (FF) dependency relationship (Figure 30) is the opposite of the start-to-start relationship in that both activities must finish at the same time. That does not mean that the two activities have to start at the same time, only that they end together. Think of making dinner as an example of this scenario. The meat, vegetables, and salad have different preparation times and therefore start at a different time. Nevertheless, they must all be ready to go on the table for dinner at the same time.

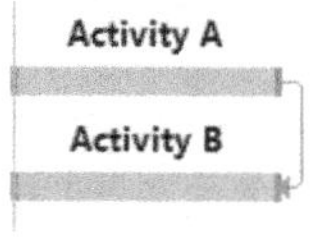

Figure 30. Finish-to-Finish (FF) dependency relationship

SDLC PHASE OR MILESTONE REVIEWS

Depending on organizational preferences, a program or project team reviews the major milestones, deliverables, activities and tasks, budgets and resource information before starting or completing a new phase of work. The use of the word phase does not necessarily equate to the SDLC phases. In the context of Agile practices, the term phase most often refers to a product release, of which there can be many over the life cycle of a product. Milestone reviews apply to both the traditional SDLC Model and Agile models.

This author has personally seen such reviews initiated either at the start of a new phase of work or at the end of a phase of work. Truly it doesn't matter, because a review of both the entry criteria for the next phase and exit criteria for the preceding phase occur in phase review meetings, and there is no reason to have separate meetings except to follow up on checklist items that did not pass the initial phase review.

The outcome of the phase or milestone review is either approval or disapproval to proceed on to the next phase. The phase review teams review and confirm the predetermined exit criteria that demonstrate completion of work in the preceding phase and conforming with the specified acceptance criteria. In contrast, the predefined entry criteria indicate the preparedness of the development team to start working on the next release of the product.

The term "phase review" is a holdover from the traditional SDLC models and does not make a lot of sense in an Agile-based project. The terminology used is not the issue. It's more important to understand the purpose of such reviews, which is to ensure preparedness before releasing a product and moving on to another work iteration.

Therefore, under Agile-based practices, conduct milestone reviews between each iterative release of the product that is shippable to a client. Also, the first milestone review should start at the end of the concept phase. This initial review confirms the organization has completed the essential work necessary to determine the required deliverables, a positive ROI, appropriate project constraints (e.g., scope, budget, time, resources, quality, and risks), and executive or client approvals to proceed.

Entry Criteria

Before starting work in a new life cycle phase, some activities and deliverables must be completed to ensure that the program or project team is ready to work on the next phase. With that in mind, it is not unusual that a corporate SDLC has predefined entry criteria and checklists that must be completed and reviewed before proceeding on to the next phase. Note that the entry criteria at the start of a new product development life cycle effort differ from succeeding entry criteria, as the initial review needs to make sure the program or project business case, approvals, and constraints are complete and validated.

Exit Criteria

Like entry criteria, this section calls out specific deliverables, activities, and tasks that must finish before the project or program team can receive authorization to close out the work in the current phase of work or release of the product. Successful review, verification, and approval of Exit criteria for the previous phase or release is required before the development team can proceed onto the next phase of work. As with entry criteria, a corporate SDLC should have predefined exit criteria and checklists that must be completed, reviewed, and approved before proceeding on to the next phase.

Phase Review Checklists

Having predefined entry and exit criteria templates eliminates the need for each product or project team to reinvent the wheel every time they complete or start a new release. The team may need to revise the templates to cover unique requirements but building the entry and exit checklist for each release from scratch makes no sense at all.

In preparation for a phase or milestone review, the project manager and project team can complete both sets of checklists in parallel. The entry and exit criteria data should be sufficiently detailed to demonstrate compliance with requirements and readiness to proceed on with the next phase of work. These checklists help ensure that the development team dots the i's and crosses the t's before seeking approval to exit one phase or release and proceed with the next phase of work or release.

CHAPTER SUMMARY

In this chapter, you have learned that a mature SDLC provides a framework to classify, organize, manage, discover, and ultimately use information related to systems development engineering and best practices across a product's life cycle. One method to do this is for the organization to define a phase-based hierarchical classification structure for managing information across a SDLC. Each phase represents a type of work common to systems development. Taxonomy is the common term used to connote hierarchical classification schemes.

Next, we turned our attention to understanding some of the common types of information elements managed within a corporate SDLC. The primary information elements within an SDLC are listed below:

- SDLC phases
- SDLC processes
- Activities
- Tasks

- **Activity worksheets**
- **Process and activity flow charts**
- **SDLC descriptions**
- **Goals and objectives**
- **Responsibility assignment matrixes (RAM)**

Next, we explored in some detail the use of an activity worksheet as a template to help initially document and improve the maturity of a defined process, activity or task.

Finally, we discussed the importance of having periodic milestone or phase reviews. One of the important tools for preparing for such periodic reviews is the phase review checklist.

The next chapter demonstrates an effective technique for integrating multiple SDLC processes and project management life cycle processes to develop custom project plans.

12. Integrating Disparate Scopes of Work

This chapter demonstrates how product breakdown structures, SDLC processes, and PMLC processes integrate to define the total scope of work required to complete an IT project.

CHAPTER OBJECTIVES

At the end of this chapter, the reader should be able to discuss the following:

- Explain how project management methods and tools support different types of programs and projects across industry, product, service and solution applications
- Explain why it often makes sense to break up a product or systems life cycle into separate projects, each having a separate PMLC
- Explain how the activities and tasks that span the project management process groups form a single PMLC
- Discuss the reasons and methods for integrating both Agile and lstructured (top down) practices in an SDLC
- Discuss the activities and tasks typically contained in a project management WBS
- Discuss the activities and tasks typically contained in a systems development WBS
- Explain how the PMLC and SDLC WBS templates can be tailored to fit the needs of an IT project

AUTHOR'S INTRODUCTION

Project managers are familiar with the concept of project integration. For example, a large project may involve work performed by several organizations operating independently but collaboratively on different parts of the overall solution.

A key role of the assigned project manager is to work with management, the project team, and other stakeholders to define the total scope of work that must be planned, coordinated, and scheduled. An integrated master schedule (IMS) is the term most often used to describe an overarching schedule that involves tracking work and resource assignments across multiple but related programs and projects. The project manager or scheduling expert builds and maintains the IMS throughout the project or program.

As an example, I managed a large Health IT project with 98 people organized into 16 discrete sub-projects, all of which had different project team assignments. I had to work with each team to identify the scope of work for each subproject and include each teams' tasks within the IMS, and then oversee execution and delivery of the work. Similarly, I managed a larger project portfolio within another Health IT program involving 115 project team members and stakeholders that required integration of team efforts supporting ICD-10 diagnostic code remediation across 32 clinical applications. In both examples, given the size of these programs, synching and coordinating project task deliverables and outputs presented significant challenges.

Programs initiated to support strategic change initiatives often have separate projects working on different pieces of the initiative, with each project team having resources, deliverables, and schedules integrated into the master schedule. There are also different types of work components that must be integrated, such as work associated with the products, work associated with project management, and work in support of other corporate processes.

In this chapter, I present an approach to integrate standard project management processes and related activities with the processes and activities associated with developing products under a SDLC. In this discussion, I'll show how this works within the traditional waterfall approach, the hybrid model, and in support of Agile-based development practices.

STANDARD PROJECT MANAGEMENT PRACTICES

Basic project management practices apply to manage all types of projects. It's a skill set in its own right—one that stands alone regardless of the industries, products, services, or other solution applications. For example, while I have spent most of my career managing IT projects, I have also managed environmental remediation and construction projects for the US Forest Service, and I have managed projects involving manufacturing equipment development, facility layout reconfigurations, and equipment implementation projects in a Fortune 500 technology firm.

From this experience, I have come to understand that the methods and tools used to manage all projects remain pretty much the same no matter what type of projects I'm managing. Business initiatives that require project management support are pervasive in every organization, particularly in the large enterprise, and they must develop mature project management processes and practices to address these needs.

The discipline of project management also has a life cycle set of processes, similar to the concepts employed under SDLCs. However, a single PMLC does not map necessarily to the entire scope of work spanning a product's life cycle.

For example, a product's life cycle may span many years and include multiple enhancements. Each new enhancement represents a unique change initiative requiring budget reviews and approvals, and therefore, project management oversight to ensure that the funded deliverables meet expectations. As a result, the organization may initiate multiple projects, each managed independently, across the lifespan of a product or system.

Multiple Project Life Cycles

The most common reason that a product has multiple projects over its lifecycle is to oversee the development of new features and functions. In the traditional SDLC model, such allocations are driven by new demands that cropped up in a previous development cycle, or after the product is released.

Under agile practices, the organization may continue development on an ongoing basis, across fiscal years, so long as the demand for new features and functions justifies the investments. Still, executives must re-evaluate their IT investment decisions on an annualized basis and approve new funding for continued development. As with the traditional SDLC model, the assigned project manager provides the oversight to keep the project on track in terms of producing the approved deliverables within the constraints imposed by the project's charter.

However, especially under the traditional SDLC model, it's also not unusual to have multiple projects authorized within a single product development cycle. In such cases, the organizations may divide the SDLC across multiple projects to reduce risks. Likewise, a product's end-of-life processes may require project management oversight when there are significant equipment disposal, data protection, and new system transition issues to consider. Moreover, on very large projects, it often makes sense to divide the work across multiple project teams, typically aggregated and managed at the program level.

Under the traditional SDLC model, an organization may decide there is too much that is unknown, and therefore too much risk to authorize funding and resources for an entire development lifecycle. So, they may break out concept, requirements, and design phase work into one or more separate projects.

The traditional waterfall model too often leads to the development of products that are not in alignment with the needs of the customers and end users. By the time the customers and end users have access to the product, the allotted schedule time and budgets are largely expended, making it nearly impossible to make any modifications to the product. To address such concerns, the IT organization may encourage the development of a prototype as a separate project, or as part of a design phase project. Funding a prototype project reduces risks of getting things wrong right from the start. Also, a prototype effort provides more data for the team to scope out the level of effort to develop the final solution before the organization makes final commitments to resources, time and budgets.

Figure 31 illustrates one concept of having multiple projects initiated across the lifespan of an IT system. In this scenario, the organization has split the concept phase work, requirements and design-related work, development and testing work, deployment activities, and post-production related work into individual projects. Each project has a separate lifecycle from a management perspective.

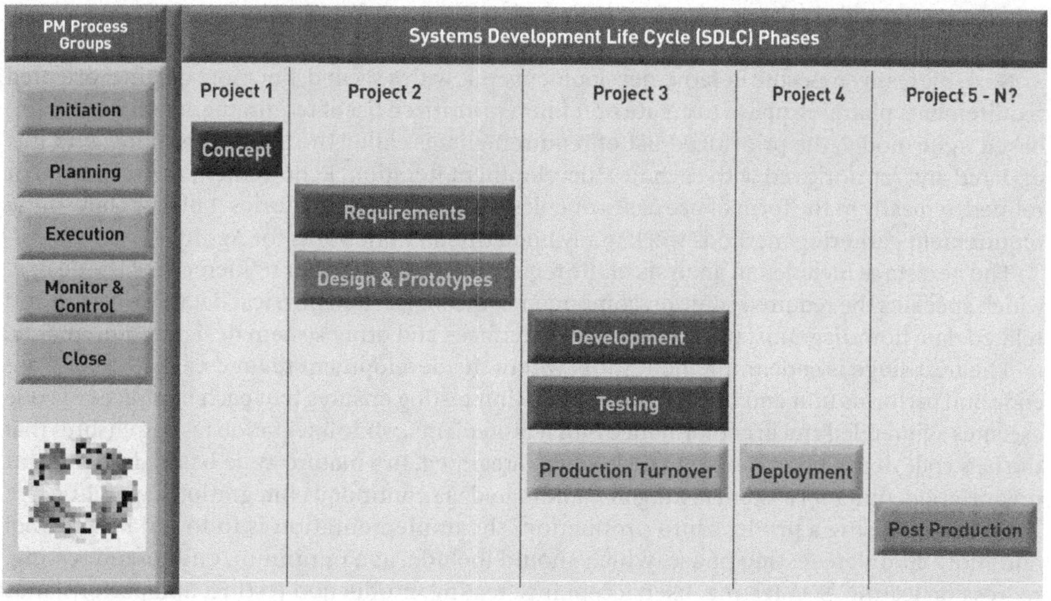

Figure 31. Splitting a Systems Development Life Cycle Across Multiple Projects

The IID philosophies of Agile help address the issues of managing risk by frequently and iteratively adding new functionality, and correcting bugs and defects. In the traditional waterfall approaches, it is not uncommon to see individual projects set up to manage the initial concept development, requirements analysis, architecture and design analysis, and another project to manage development and testing activities. Breaking a product life cycle across these individual phases helps reduce the risk of building a product that is not initially well defined.

During my career, I have managed projects where the entire authorized scope of work only included the development of a CONOPS Document and high-level Business Requirements Document (BRD) – both of which logically fit within the concept phase of an SDLC. I have managed other project assignments where the total authorized scope of work included defining user requirements and sometimes included the system's design. Also, I have managed projects where the scope of work included only the development and review of a prototype or COTS application.

Project Management in Support of Agile-based Development

Under Agile, as noted previously, projects tend to be set up to support an approved scope of work across a fiscal year or to complete a very specific scope of work. Like projects established to support annual waterfall-based releases, project assignments for annualized Agile-based projects gives management and customers increased confidence that the team remains focused on meeting the business requirements that justified the investments. The periodic project milestone reviews and status updates provide the data executives and customers need to ensure that the development of the final solution is proceeding within the approved constraints.

Most IT developers have come to understand that a truly linear development life cycle process isn't efficient or effective at producing customer-centric and truly useful software products. The Agile life cycle process graphic, shown previously in Chapter 8, Figure 19, is more representative of how life cycle phases are cyclical, following the IID development pattern. In other words, this graphical view of Agile practices implies that some number of separate life cycle iterations may be required to develop an entire product, system, or solution by adding value incrementally over time.

In the iterative, Agile model, an initial planning phase is required to obtain a baseline set of requirements, as well as obtain customer commitments, and validate existing project constraints. During the initial planning phase, the team must estimate the scope of work they can complete within the time, resources, budget, and quality constraints imposed by the client. It's also prudent to begin working on the high-level solution design and software architecture at this point, as initial choices made in the architecture become foundational for the solution, and increasingly more difficult, time-consuming, and costly to change with each new iteration.

Next, the team enters the iterative development cycle with a second, but more detailed-oriented requirements planning phase to create or refine a prioritized list of requirements. In the Scrum-based Agile model, the prioritized list of requirements is called the product backlog, and it is updated and reprioritized with each new development iteration. Requirements are captured or refined, typically in the form of use cases or epics, themes, and user stories. I discuss how these requirement gathering methods work in my book titled Frameworks for Agility.

The next stage includes an analysis of the requirements and further refinement of the design, which specifies the required solution components, the logical and physical database design and related data flow diagrams, plus component interfaces and other system design requirements.

The next stage is code implementation, where the development team creates the software code and performs unit and integration testing. Unit testing ensures that each new piece of code executes as intended and in conformance with requirements, while integration testing ensures that the new code doesn't break existing code when integrated. In a mature Agile-based development organization, unit and system testing are automated via continuous integration capabilities.

Before releasing a product into production, the implementation is followed by a formal and more complete testing phase, which should include, at a minimum, end-to-end testing, regression testing, SQA testing, user acceptance testing, production testing, and performance testing. I discuss the purpose of these testing types in my *Frameworks for Agility* book.

If the code produced is ready for production release, the development team deploys the product or system to the customer or end user community. If not, customers and end users evaluate the incremental releases to determine if there is a need to modify or add new requirements to the product backlog.

The evaluation phase of this generic Agile cycle includes a customer or end user review to demo the product and get input on new enhancement ideas and backlog priorities. In the Scrum methodology, the team also reviews the work they accomplished to look for things they can improve on. At that point, the Agile cycle begins anew for the next iteration.

INCORPORATING AGILE CONCEPTS WITHIN A STANDARD SDLC MODEL

In chapters 9 and 10, the SDLC models depicted follow the traditional, top-down pattern, although we loosened up on the concept of linear, sequential progressions by showing overlapping SDLC phases. However, a modern SDLC model must also incorporate prototyping and IID techniques, which are the hallmarks of an Agile-based development process.

The graphical SDLC model shown in Figure 32 depicts Agile IID practices in context with traditional life cycle phases. Each build represents a potentially shippable product. What this figure is showing is that each build of a product has a separate life cycle, consisting of activities from requirements through postproduction, though the later tasks are typically handed off to an IT Operations group.

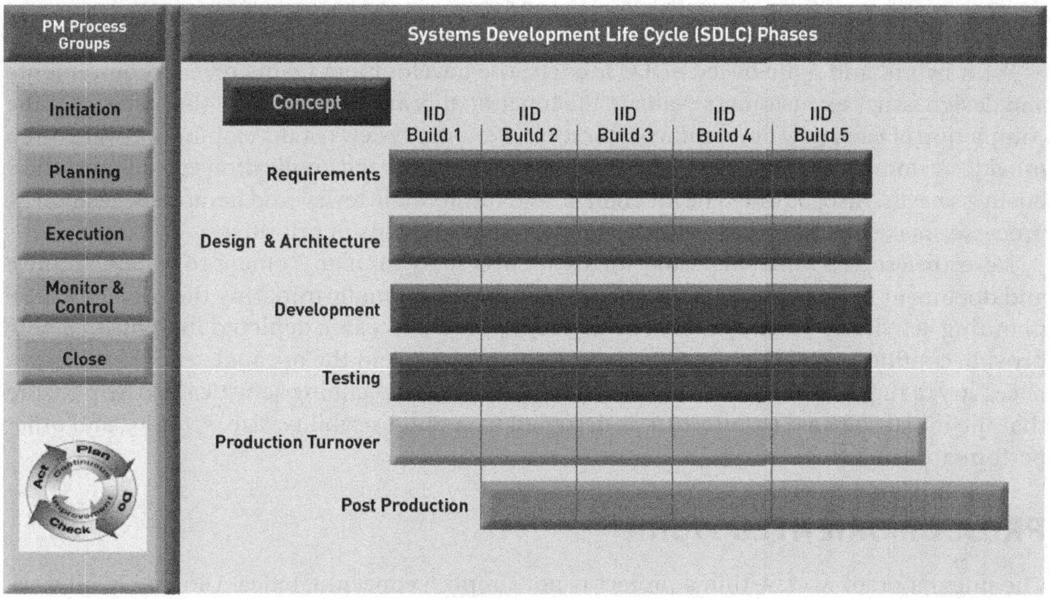

Figure 32 SDLC Phases Spanning Development Iterations and Releases

Again, as discussed in the previous slide, it can make sense to conduct initial high-level requirements and design assessments before starting development, to refine the product's concept and build the business case to justify the resources and expenditures, even when implementing Agile practices. In the traditional phase based SDLC model, these activities occur in the concept phase.

Setting up the initial high-level requirements and design assessment as a separate project gives the project team time to assess all business, customer, user and design requirements necessary to scope the project properly. It also gives management and customers a chance to review the initial assessments before making final development investment decisions and budget commitments.

One final note. In modern Agile development shops, particularly in support of large-scale enterprise or commercial software applications, it's becoming more routine to have continuing development iterations across the entire lifespan of the application. In other words, every few weeks or even shorter periods, a new release is made available as a "potentially shippable product." However, even then, it still makes sense for executive management to charter, as a separate project, high-level requirements and design assessment activities, to set proper product development expectations, budgets, and other constraints for the upcoming fiscal year.

BENEFITS OF CONDUCTING AN INITIAL CONCEPT PHASE

As noted in Chapter 2, waterfall is not completely dead—yet. Plus, as I noted, there is nothing to state that waterfall practices can't implement prototyping and IID development practices. Recall Herbert Bennington states that this is exactly how they implemented their structured SDLC practices on the SAGE program, back in the mid-1950s. I also noted that Structured SDLCs typically have staggered and overlapping phases, based on activity level dependencies. In Figure 33, the SDLC phase overlaps show this concept graphically across each development iteration and release.

The waterfall model remains attractive to some customers who cannot or will not devote the time or resources necessary to support frequent reviews throughout the development life cycle. It is hard to get executives, customers, and end users to commit to requirements-gathering sessions at the start of a waterfall-based project. It's even more of a challenge to obtain time, budget and resources to conduct a concept phase level of review of requirements to determine the business case and ROI, plus establish basic business and user requirements. Sad, but true.

However, it is not a good practice to have the project team and developers try to figure out the requirements on their own without initial and ongoing access to executives, customers, and end users for direction and feedback. That is pretty much a sure path to failure.

With hybrid and Agile-based SDLC models, the development teams revisit requirements and design issues as customers evaluate the functionality and usefulness of the software at the completion of each new development iteration. Periodic reviews of developing software often provide customers and end users with new insights on how the application can support their business needs. Also, business needs change, and the periodic review and iterative development processes make it easier to adjust to a change in requirements or priorities.

Nevertheless, the initial planning stages are necessary for management to review, approve and document the project team's scope or work before formally initiating the project. These planning activities, which include the concept phase processes depicted in Figure 33, help provide confidence that the investments in the project yield the optimal set of capabilities necessary to support the mission of the business. The initial planning activities also help ensure that the initial software architecture is designed to support usability, size, security, and other performance objectives.

PROJECT-ORIENTED WORK

The integration of work within a project is not simply a conceptual idea. Ultimately, all work must be planned, scheduled, executed, monitored and controlled, and closed out. However, before we can do any of that, we first must define the scope of work for the project.

In an IT project, work consists of a combination of project management activities and systems development activities. However, standard SDLC processes lack details on the specific requirements of products under development within a project. Therefore, project teams must also discover and integrate the work involved in fabricating and assembling parts into the final product or system.

Scoping work almost always involves breaking a hierarchical decomposition of higher-level deliverables into smaller components until we get to a listing of the lowest level parts that make up the process, product or system. Hierarchical decompositions are called breakdown structures. When a breakdown structure decomposes a product, service, or some other desired outcome to identify the lowest level activities and tasks, we refer to the decomposition as a WBS. (see next subsection titled WBS Overview)

A WBS is the initial tool project managers use to define the total scope of work. Depending on the size and complexity of a project, they may be a need to define separate work breakdown structures for different components of the work. Most project schedules should include and integrate at least three types of work decompositions to create the project schedule encompassing a product WBS, development activities, and management activities.

The most obvious breakdown structure is the product's WBS, which can be thought of as an exploded parts list, with each component having some sort of work associated with it. However,

SDLC and PMLC processes also have deliverables-based decompositions that should be integrated with the WBS to define the overall scope of work. Before we get into that subject, let's discuss in more detail what a WBS is and how it is used.

WBS OVERVIEW

A WBS is a deliverable based decomposition of processes, activities, and tasks. For example, the WBS describing a PMLC starts with a listing of typical project management process groups or phases. The individual processes must identify one or more deliverable that is precisely defined and measurable.

Processes decompose into lists of the activities and tasks required to create the higher-level project management deliverables. However, note that activities and tasks must also have definable, specific, and measurable deliverables. Otherwise, there is no way to know whether the work is complete.

A WBS decomposition continues along each of the product's component branches until further decomposition cannot yield a specific and measurable result, or until the level of work is too small to justify management oversight. (Recall the 8/80 rule discussed in chapter 11.)

The same hierarchical and deliverables-based decomposition concept applies when developing a WBS for an SDLC process. The project team starts with the SDLC phases, then decomposes the processes into activities and tasks, with each level representing smaller pieces of the work, and each having specific and measurable deliverables in the form of products, services or desired outcomes.

WBS Naming Conventions

When defining a product or process oriented WBS, a useful concept is to think of the highest-level deliverables as answering "what is produced by the project" in the form of products, services, or desired outcomes. Therefore, as a good practice, the highest-level deliverables specified within a WBS should have names expressed as nouns or noun phrases to indicate they are tangible or intangible items, such as "unit testing process."

The lowest levels of decomposition are used to define the work required. In other words, the focus of the WBS turns from "what" is to be delivered to "how" the team creates the deliverables. Specifically, activities and tasks describe how the development team will create the higher-level deliverables. Since activities and tasks imply the notion of work or action taken, a best practice approach is to name activities and tasks with verbs or verb phrases, such as "conduct unit testing," or "develop test scripts."

So, in summary, higher-level WBS items should have names expressed as nouns or noun phrases that represent the deliverable items of the project, while lower-level WBS items should have action-oriented verbs or verb phrase-based names as a convention to designate the work performed via activities and tasks.

Do you have to name WBS component items this way? No, but it does make it easier to understand the intent of each item. For example, is the item representing a deliverable component, a work activity, or a task? The discipline of starting a WBS by identifying and decomposing the deliverable items into work items keeps the focus initially on identifying the items that must be produced by the project team, but not at the sacrifice of defining the work associated with each deliverable.

Defining a Deliverables-Based WBS

When reviewing or making a WBS, it's important to remember that a WBS is not a process model; rather, it is a deliverables-based decomposition of component work items or artifacts that have been identified and authorized as belonging to the scope of the project. As with any breakdown structure, a WBS provides a listing of related sub-elements from the highest level to lowest level subjects.

A WBS does not provide information on the flow of work. Information on workflow comes from the process, activity, and task descriptions, and specifically their predecessor and successor dependencies. The process, activity, and task descriptions also include detailed information on required resources, materials, tools, standards, estimated durations, quality metrics or acceptance criteria, and other critical information necessary to schedule and complete the work.

When planning a project, it is important to remember that the purpose of defining the WBS is to specifically ask ourselves what deliverables are required by the customer and management, and then further decompose those elements into ever smaller units of work that can still have a definable deliverable component. Specific activities and tasks are defined only once for each deliverable's decomposition.

As noted previously, dependency relationships define the flow of work. Other models and methods, such as activity network diagrams or Gannt Charts, are necessary to visualize the flow of work defined within a WBS. Ideally, each activity, task, or work package has one or more activity worksheets associated with them that provides the dependency relationships. Also, the activity worksheets may define detailed steps to complete the work. The detailed steps are found in the Procedures or Methods fields.

In some cases, the documented steps may indicate repetitive work. As an example, a software development team working under Agile-based practices doesn't create and test code once; instead, they may develop their code through trial and error in incremental steps, followed by unit and system integration testing in every code development iteration until the code works properly.

Within the WBS, it's doubtful that a task called "develop and test code" would be listed, even though there is a deliverable component to such work. However, a WBS is deliverables based—meaning the capability implemented by the code is what's critical, not how the work is performed. The capability requirements for code are contained in user stories or other forms of requirements documentation, and the work packages associated with the user story point to the "develop and test code" task.

User stories are short, simple descriptions of a desired product capability, told from the point of view an end-user or customer. In addition, user stories are the smallest level of user or systems requirements and are documented and implemented as a specific feature or functionality within the software.

Moreover, the repetitive coding and testing work does not show up in the project schedule, as the work is probably too short in duration to comply with the 8/80 rule. Therefore, only the user story or feature requirement should be recorded within the WBS. The repetitive steps required to develop code, perform unit testing, and perform system integration testing are documented in the Procedure/Method fields in the appropriate activity worksheet(s).

It is also important to note that each node in a WBS represents a unique and standalone deliverable. In a software development project, the deliverables are typically related to functional and system requirements that must be coded within the application. More complex systems have WBS decompositions that can include computer security, network, hardware and software components, plus other electronic and mechanical components.

The hierarchical separation of WBS nodes by component deliverables is necessary as the decompositions help expose the total scope of work required for a project. As I've noted before, within the WBS, I prefer to define activities as a collection of work around major customer-oriented deliverables. Since the deliverables represent tangible or intangible things created by the project, it's not unusual to see activities described by nouns and noun phrases. Tasks define work as an action, and task names should therefore be expressed as verbs or verb phrases.

Within a WBS decomposition, the assignment of tasks as work occurs at any level. I've found that many, if not most, descriptions of how a WBS works do not always make this point clear. Too often, the words "deliverables" and "work" are used interchangeably. They are not the same thing. Deliverables are the artifacts created or enhanced in a project; work is a procedure or method for creating the deliverables. Deliverables are defined by the requirements, while work is an approach to creating or extending a project's deliverables as defined by the organization's SDLC processes.

Figure 33 shows an IT system as a solution that includes both hardware and software components. The hardware and software components are deliverables. A complete WBS shows all the deliverables associated with software and hardware as separate branches in the hierarchical decomposition to whatever level is required. In Figure 33, the software decomposition includes two modules, while the hardware decomposition includes servers and networks. In a classic

WBS, the work descriptions are never shown in the hierarchy, just the list of deliverables. Even the work packages are mapped to deliverable items.

As an example of work, software is installed and configured on hardware components. The organization's SDLC processes describe the configuration and installation processes. A work package might be established at that level to indicate the configured and installed software as a deliverable. But that's not the lowest level of our WBS. As shown in Figure 33, the software branch of the WBS has further decompositions that indicate development requirements for two independent software modules. Each of those deliverables also has work associated as tasks, such as requirements and design reviews, coding, unit testing, and integration testing.

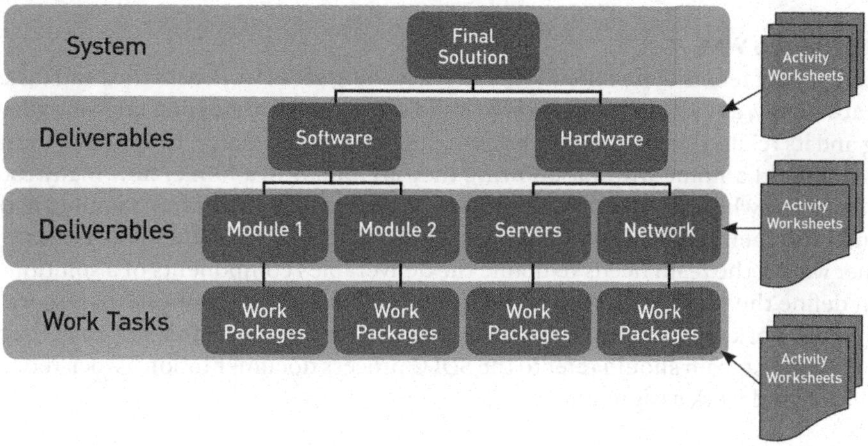

Figure 33. IT System WBS Example

Likewise, the hardware branch may require acquisition, installation, and configuration of components and subassemblies. All of this work is documented as procedures within the organization's SDLC processes and related activity worksheets. Rather than define deliverables for each of these work activities, it's often easier to just point to the appropriate SDLC processes and related task descriptions that define the procedures for performing the work associated with each deliverable.

Using the Scrum method as another IT example of how to build an IT-based WBS, work is typically defined in themes, epics, and user stories. Themes are useful collections of user stories, perhaps around a high-level functional requirement. Epics are essentially large user stories expressing multiple requirements that need to be decomposed and refined in the form of multiple user stories.

The project manager can use both themes and epics for the higher-level deliverable components within a WBS, while the lowest levels of a WBS are the associated user stories. As a quick side note, since by definition the details of a project are progressively elaborated over time, the initial WBS and project schedules may only include a listing of epics and themes at the start of a project, serving as placeholders for future sprints and releases. As the user stories are refined and prioritized within the product backlog, the WBS and project schedules should be updated.

As the user stories evolve, they specify the deliverables of the project regarding required features and functions. However, the explicit procedures describing how to produce the features required by each user story are documented in the organization's SDLC processes. The SDLC processes are mapped by the SMEs to the deliverables as part of the ongoing project planning effort. The SDLC process and related activities and tasks have the same form as the activity worksheet templates.

How to Build a Proper WBS

The process of building a WBS starts by asking ourselves what we are creating. In other words, we need to define the highest-level deliverables, and then ask ourselves what components, or subassemblies or other artifacts are required to form the higher-level deliverables. In turn, those components and subassemblies are further decomposed to expose and describe lower-level work in increasingly smaller, yet manageable tasks.

Once the smallest level practicable deliverables and components are defined (work packages), we then ask ourselves how the team will create those deliverables. The answer to this will define the work necessary to create the deliverables. The SMEs assign SDLC and other corporate processes that contain the GEP and best practices for the work associated with each deliverable.

The number of levels of work described within a WBS varies from project to project. Also, the number of decompositions within each branch of a WBS can vary. The number of decompositions is not as important as defining the work with enough granularity to achieve 100% of the work required to support the approved charter, and not 1% more.

How Not to Build a WBS

I have seen project teams define their WBS as a cascading set of work activities and tasks. That's not a great strategy, as you can't know how to build something without first knowing what you are building and its related subcomponents. By analogy, a builder wouldn't simply list the activities and tasks to build a home without knowing they are building a home, how many rooms and bathrooms it has, what electrical and plumbing systems and materials are required within each room, and what their budgets and schedules are. IT developers face the same issues.

In other words, the team needs to define the deliverables components of a solution initially, and then define the work necessary to produce each deliverable. It's never necessary to show the individual work tasks within the WBS, activity network diagrams, or project schedules. Instead, the project team should refer to the SDLC process documentation associated with their deliverables-based work assignments.

INTEGRATING PROJECT MANAGEMENT AND SDLCS

So far, we have not discussed how project management processes fit within a standard SDLC model. Let's do so now. As with the systems development processes, standard project management processes guide the PM and team in the use of proven practices, activities, tasks, methods, and tools as aids in managing project-based work. PM processes are completely separate activities from the activities listed in the corporate SDLC framework.

I noted previously that there could be any number of projects initiated across the life cycle of a product. As shown back in Figure 33, we can see that the project management processes stand alone, outside the SDLC phases, and are included on every project initiated in support of developing or sustaining the product or system. Again, there can be any number of projects initiated across the life cycle of a product or system.

How we include PMLC processes in a project schedule is no different than how we handle SDLC phase-based processes required within a project. In both cases, we list the project management deliverables in the project's WBS, and the PM refers to the related processes, activities, and task descriptions to ensure that they are following the organization's standard and best PM practices.

A method for integrating PMLC/SDLC deliverables is shown later in this chapter. However, first, we are going to dive a bit deeper into the difference between PMLC and SDLC processes.

PROJECT MANAGEMENT PROCESSES

The discipline of project management provides standardized processes that are proven to aid in the successful delivery of project-based deliverables within the allotted constraints approved by executive management (or other paying customers). Similar in concept, both systems development and project management processes provide a set of best practices that guide practitioners in their work. Standardized processes and best practices help improve quality, efficiencies, and consistency throughout the organization.

All processes, activities and tasks have deliverables and acceptance criteria associated with them. If that were not the case, there would be no way to know if the work is executed properly. Processes are an expression of work as defined by a set of activities and tasks. The output of the process is a final deliverable or set of deliverables.

A project management process to develop the schedule for a project has a related deliverable of the project schedule. In other words, the scheduling process is a description of work, described through a series of activities and tasks, while the deliverable of the scheduling activity is the project schedule.

Just as products have a life cycle, so do projects. In other words, both products and projects have a beginning, an ending, and something going on in between, guided by standardized processes and best practices. The goal of having life cycle processes is to reduce risks and improve the consistency of outcomes.

Project Management Life Cycle (PMLC)

Conceptually, the management of projects follows a repeatable set of life cycle processes on every project, regardless of the scope of work contemplated. The basic PM life cycle processes follow the pattern described below:

- Define the work
- Plan the work
- Start the work
- Manage the work
- Finish the work

The application of project management practices is most commonly associated with managing work under the traditional waterfall based SDLC model. However, any type of activity that develops a unique product under time, quality, and budget constraints benefits from the application of project management practices, including IT projects managed under prototyping, IID, Hybrid, Agile, and lean SDLC models.

The primary point of this section is to note that the discipline of professional project management includes standardized processes and best practices that, as part of the project planning process, project managers must integrate with the organization's SDLC processes. Where the SDLC process defines how to perform the work, project management processes define how to manage the work.

SDLC versus PMLC

By now, it should be clear that the two life cycles, PMLC and SDLC, are very different. The focus of the SDLC is on deploying a set of processes and best practices to design, create, support, and sustain complex and unique products, systems, or solutions. In contrast, the focus of PMLC processes is to deploy a set of processes and best practices to manage project-oriented work.

Throughout the life of a product or system, the SDLC processes guide the development and support teams' activities. In parallel, the PMLC processes guide the work of project managers to ensure that the deliverables are produced within the authorized constraints of the project. Project constraints include the approved scope of work, costs/budgets, time/schedules, and quality/acceptance criteria.

Project management processes provide a set of best practices to minimize risks and scope creep, and to manage the impact of change that affects the approved scope of work. The project manager also ensures that project team members adequately plan the scope of work required to create the deliverables.

Here's something to think about: IT project management practitioners have traditionally supported changing requirements through a rigorous change management process. In contrast, Agile-based SDLC practices support changes in requirements with little formal approval process outside periodic customer and end user review and the approval by the product manager. So, while the change approval process has changed, it's still the role of the project manager to oversee the work to ensure conformity with the project's charter and approved constraints.

Key Personnel Roles

My book titled *Frameworks for Agility* gets into quite a bit of detail on roles and responsibilities across the SDLC processes. However, I want to spend a moment here discussing the roles of product and project managers, as it may seem they have overlapping responsibilities, particularly in Agile-based projects.

Having multiple managers overseeing work seems like a recipe for disaster. Especially when the two managers ultimately support the same products and report to the same executive managers. However, as it turns out, they are overseeing different aspects of the project.

The role of the product manager is to ensure that the organization is creating and delivering the right set of products in support of the strategic goals and objectives of the business. In contrast, the role of the project manager is to ensure that the work is properly planned and executed within the approved constraints of budget, schedule, quality, and resources. In summary, and as their titles imply, product managers focus their efforts on planning and delivering the organization's products, while project managers plan and deliver the organization's authorized work across projects.

Another role that might be confusing is that of the Scrum Master. The Scrum Master is an assigned role on Agile projects, such as Scrum, XP, Kanban, and SAFe®. They too have duties that sound like those of the project manager, such as meeting facilitation, problem solving, team building, and communications.

A Scrum master acts more like a coach to ensure that the team is properly following the Agile practices, and to eliminate issues or other impediments that prevent the team from doing its work. By contrast, the project manager represents the interests of the executive sponsors or other paying customers. Both individuals have a clear interest in the success of a development project.

The real difference across all three roles is the focus of their efforts. The product owner's focus is on supporting the product. The Scrum master's focus is on supporting the team. Moreover, the project managers support those who are paying for the work and other stakeholders who can negatively or positively impact the results of the project.

Table 16 defines the typical roles and responsibilities of product and project managers.

Table 16. Roles of Product Managers versus Project Managers

Product Manager	Project Manager
Defines the product strategy and development priorities in alignment with corporate strategies	Assists with building the project team, and onboarding and offboarding team members
Responsible for determining product requirements, pricing, and timetables	Develops and maintains the project plan in conformance with the project's charter
Assesses the market opportunity and competition	Facilitates and is responsible for all aspects of the project planning and scheduling process
Facilitates the development of marketing strategies	Oversees risk identification, mitigation strategies, and contingency planning
Ultimately responsible for the success of the product over its life cycle	Monitors the work against the project plan and schedules, and facilitates necessary corrections
Facilitates cross-organizational collaboration to define, produce, promote, sell/offer, and deliver the product	Oversees project's change management processes
Responsible for revenue streams associated with the product	Reports progress against the established plans, schedules, and milestones
Seeks new market and revenue opportunities	Closes out project activities upon completion or termination

Project Management Artifacts

As we dive deeper into the project management processes, I will introduce the typical types of activities performed in each of the process groups. For now, borrowing the concepts of a deliverables based WBS, we can use the artifacts produced by the project management processes as deliverables of the project management processes and related activities.

Table 17 provides a list of common artifacts (documents, tools, forms, and plans) used by project managers, presented in context with PMI's standard project management process group.

Chapter 9, in my book SDLC Tools and Templates, includes project management templates that support the development of the project management activities and deliverables.

As with SDLC artifacts and templates, not every project management artifact or template is required throughout a single PMLC. However, in the long run, project managers will find most of these artifacts and the related templates, presented in my book titled SDLC Tools and Templates, useful at one time or another as they help facilitate the proper execution of the project management processes.

Table 17. List of Project Management Artifacts by PM process area

Project Management Life cycle Process group	Project Management Artifacts/ Deliverables
Project Initiation	**Project Charter**
1. Define the Work	Project Resource Onboarding Checklist
Project Planning	**Project Plan (Typical Plan Sections)**
2. Plan the Work	• Approved Project Scope/Deliverables
	• Technical and Management Approach
	• Initial Project Estimations
	• Resource Plan
	• Schedule Plan
	• Data Management Plan
	• Risk Management Plan
	• Measurements & Analysis Plan
	• Communications Plan
	• Project Reporting Plan
	• Supplier Agreement Management Plan
	• Change and CM Plan
	• Process & Product Quality Assurance Plan
	• Finance/Budget Plan
	Activity Worksheets
Project Execution	**Project Plan (as updated)**
3. Start the Work	Project Schedule
Project Monitoring & Control	**Project Status Report**
4. Manage the Work	Change Request Form & Change Request Log
	Risk Identification Form (RIF) and Risk Log
	Action Item Log
Project Close	**Lessons Learned**
5. Finish the work	Resource Off-boarding Checklist
	Certification of Completion and Compliance Form

INTEGRATING PROJECT AND PRODUCT LIFE CYCLES

A project's schedule can have varying amounts of work described, from project to project, based upon the number and type of deliverables required, the development approach, and the technologies employed to develop the solution. Similarly, not every project needs the same level of the project management process and task requirements, depending on complexity, size, and management oversight requirements for each project. The project manager must identify and then integrate those PMLC activities and tasks necessary to support the development activities without making the management oversight overly burdensome.

As noted in the previous section, the concept of using a WBS to describe the overall scope of work applies to both PMLC and SDLC processes. There can be other corporate processes that must be incorporated in a project plan and project schedule. For example, there may be onboarding processes required to bring employees and contractors over to the project team. The project plan and schedule may need to incorporate processes that govern BPI activities across the enterprise.

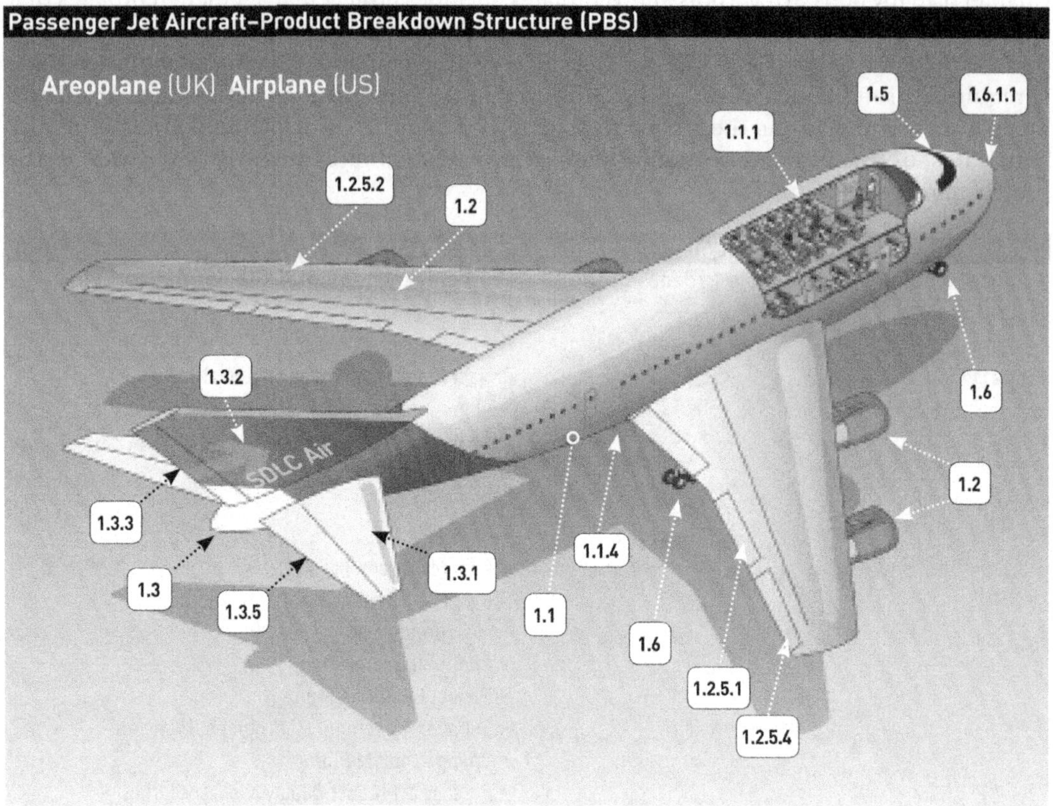

Figure 34. Aircraft Parts Example

1.1. Fuselage
- 1.1.1. Passenger Cabin
- 1.1.1.1. Seats
- 1.1.1.2. HVAC
- 1.1.1.3. Galley
- 1.1.1.4. Safety Systems
- 1.1.2. Fuel Tanks
- 1.1.3. Pressurization Systems
- 1.1.4. Hold
- 1.1.4.1. Bulk Storage Systems

1.2. Wings
- 1.2.1. Auxiliary Power Unit (APU)
- 1.2.2. Hydraulic Systems
- 1.2.3. Electrical Systems
- 1.2.4. Fuel Systems
- 1.2.5. Control Systems
- 1.2.5.1. Flaps
- 1.2.5.2. Slats
- 1.2.5.3. Spoilers
- 1.2.5.4. Ailerons
- 1.2.5.5. Trim

1.3. Empennage (Tail)
- 1.3.1. Vertical Stabilizer
- 1.3.2. Fin
- 1.3.3. Rudder
- 1.3.4. Horizontal Stabilizer
- 1.3.5. Elevator

1.4. Propulsion Systems
- 1.4.1. Jet Engines
- 1.4.2. Thrust Reversers

1.5. Flight Deck/ Cockpit
- 1.5.1. Avionics Systems
- 1.5.1.1. Communications Systems
- 1.5.1.2. Radar Systems
- 1.5.1.3. Navigation Systems
- 1.5.1.4. Information Systems
- 1.5.2. Control Systems

1.6. Undercarriage
- 1.6.1. Landing Gear
- 1.6.1.1. Nose Gear
- 1.6.1.2. Main Gear

Figure 35. Passenger Jet Aircraft Exploded Parts List.

Product-Oriented Breakdown Structures

In some ways, it is easiest to imagine the WBS as an exploded parts list. For example, we can start with a major end item, such as an aircraft, and then decompose the aircraft into its major structural parts, such as elevators, engines, fuselage, nose assembly, tail, and wings. (See Figure 34 and Figure 35.)

Next, we can decompose the major parts of the aircraft. For example, the wings decompose into smaller components and sub-assemblies such as the frame, skin, trim, aileron, and flap assemblies, fuel and hydraulic lines, wiring harnesses, fuel tanks, spoilers, and speed brakes. These components are likewise decomposed into their lower level parts lists until we get to the lowest level end items in the aircraft. The sum of these lower level parts represents the total makeup of the aircraft, and the scope of work involved to assemble the entire airplane.

When developing a product, the project team develops a WBS that mirrors the parts lists in our aircraft example. Given the size and complexity of building a modern aircraft, multiple projects are initiated to conduct the work, with many projects going on in parallel. The individual projects break the work up into manageable pieces. Completing the authorized scope of work might be possible under one project, but why make things harder than they need to be?

Software or Systems Product Breakdowns

The previous section used an aircraft decomposition to describe a product breakdown and used the analogy of an exploded parts list as an example of a deliverables-based decomposition. The scope of work then includes all the components that have to be fabricated and assembled. However, you may be wondering how that applies to software applications and systems?

There's not as much difference as you might think. Instead of an aircraft, our IT project supports the development of a software application or system as the highest-level deliverable item. The project managers can display WBS decompositions by artifacts or by features or by software modules/components. Figure 36 shows all three types of decompositions.

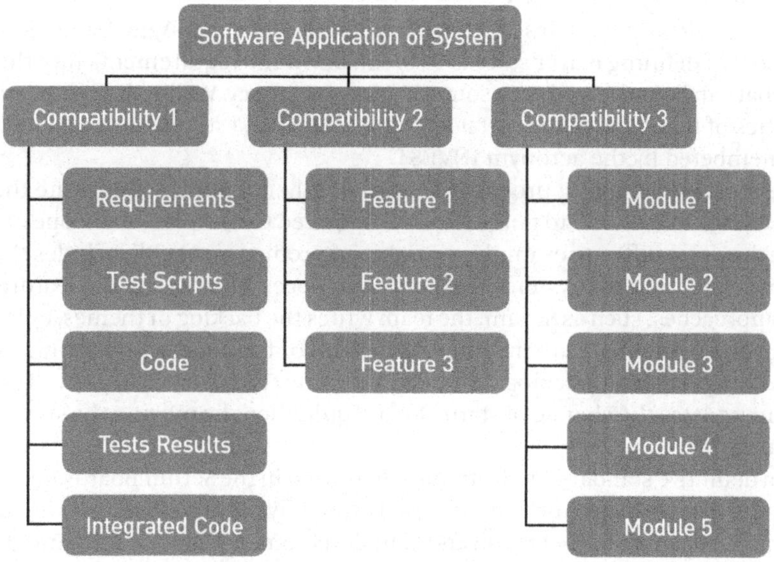

Figure 36. Example Software Application or System Product Breakdown Structure

Agile-based WBS

An organization's standard SDLC processes have a WBS structure that typically follows a taxonomy of phases, process, activities, and tasks. We showed such a decomposition in Figure 26 in the chapter 11 and in Figure 2 within Chapter 1.

There's another format that might be considered for an SDLC-based WBS, and that is to use Scrum-based requirements as the deliverables, in the form of themes, epics, and user stories. My book titled *Frameworks for Agility* provides additional details on the use of themes, epics, and user stories. For our discussion here, themes, epics, and user stories represent a potential form of product breakdown structure or decomposition, as shown in Figure 37.

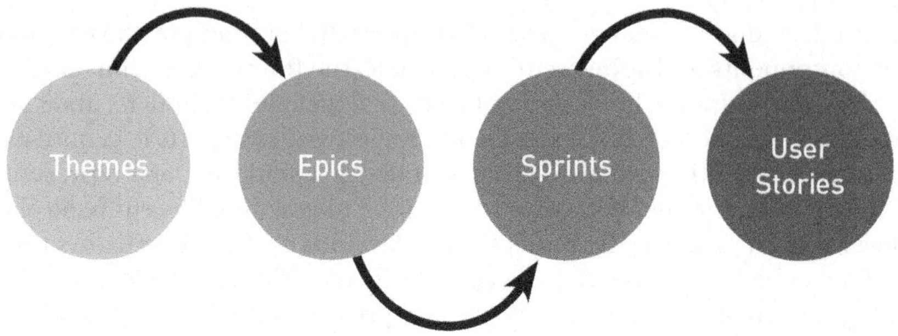

Figure 37. Decomposition of work under Agile practices

Themes are collections of high-level requirements that usefully go together, often developed together with a release or project. A theme represents a focus of work by the development team, such as the development of a system level requirement or a high-level user requirement. The main point of a theme is to aggregate work that is most efficiently completed as a group within a release.

Epics are groups of related user stories. Development teams might create epics to organize and manage related user stories within a product backlog. However, traditionally, epics allowed business analysts and developers to document requirements as a stream of consciousness from business process owners, domain experts, and end users, without interrupting to structure the requirements as individual user stories.

User stories are the lowest level description of requirements under Agile-based development practices. The goal of defining a user story is to break down the requirements into the smallest pieces of work that can be estimated with some degree of accuracy. Well-written user stories have the characteristics of being independent, negotiable, valuable, estimable, small, and testable, more easily remembered by the acronym INVEST.

One of the first tasks taken by a project manager and their team is to integrate the product oriented WBS with the SDLC WBS to guide the work required to complete the scope of work. For example, in the previous software example, a combined product and WBS includes the features required by the product and the systems development processes to create the features.

Under agile approaches, such as Scrum, the team writes the backlog of themes, epics and user stories on sticky notes, and places them on a white board or flip chart with columns indicating status under headings such as 'Backlog', 'Hotfix/ Emergency',' To Do', 'WIP', 'QA / Testing', and 'Done.' The team updates the change of status for a requirement by moving the sticky note into the proper column.

There is not a definitive standard on the terminology used in the Scrum Boards. Also, KANBAN Boards employ similar verbiage and techniques. Ultimately, use the WIP tracking terms and approach that make the most sense for you and your development teams. See Figure 38 – Sample Scrum Board, as an example.

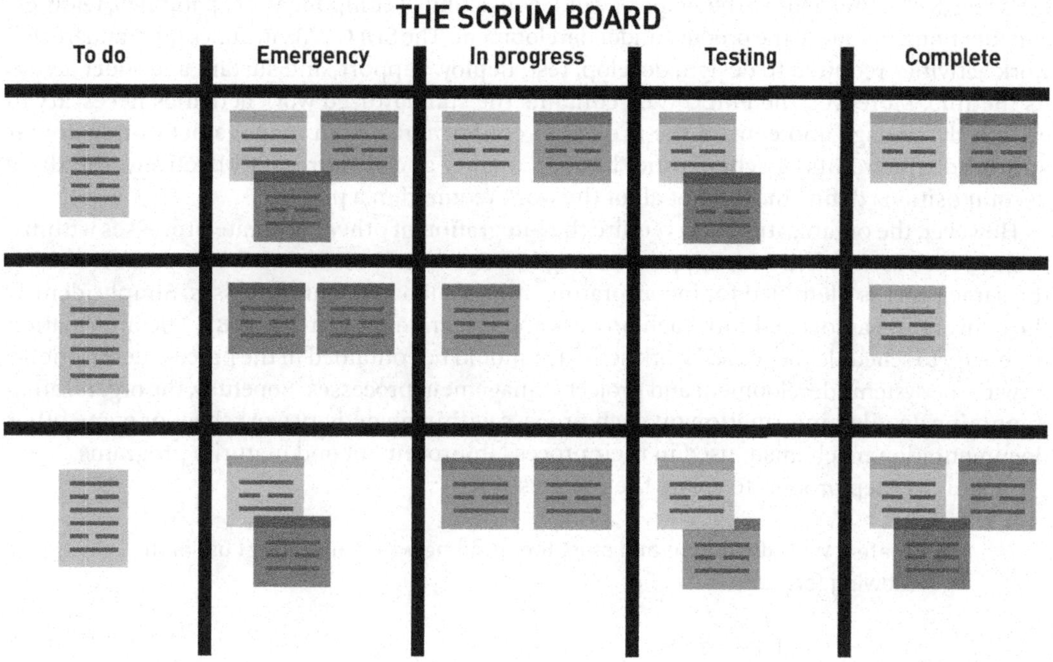

Figure 38. Sample Scrum Board

I have seen project teams implement Agile-based decompositions in their project WBS and schedule plans, and there is no reason not to include them - when your customers require such planning documentation. But it does go against the philosophy of agile, which encourages ongoing flexibility in the identification, definition and scheduling of requirements, as opposed to fixing requirements through detailed initial planning and scheduling activities.

There is a work around. What typically happens, and from a practicable standpoint should happen, is the team's project manager continuously updates their schedule plans in real time with the currently identified backlog of themes, epics and stories. So, instead of being a long-term planning tool, the schedule plan and related Gantt charts now become more of a visual aid showing short term development objectives.

The Scrum/ KANBAN Boards work just fine as visual aids for local viewing. However, the electronic scheduling tools and Scrum and KANBAN board applications enable viewing across geographically dispersed locations.

PMLC WBSs

The product and SDLC decompositions specify the work required to develop a product. Together, they typically form the bulk of the work performed under the project. However, the project manager also needs to include project management related work within the schedule.

As with the SDLC decompositions, the project manager identifies and integrates the project management deliverables associated with the organization's project management processes and their related activities and tasks within the project's WBS. Table 17, in the next section, displays a PMLC WBS Template as an example of potential project management deliverables included within a project's WBS.

Integrating Project Work

We now have identified three breakdown structures for a project, to include:

- **Product breakdown structure (PBS)**
- **SDLC WBS (SDLC WBS)**
- **PMLC WBS (PMLC WBS)**

Let's recap. The PBS defines the scope of work, consisting of components, capabilities, features, and functions, unique to the product under development. The SDLC WBS includes the standardized work activities required to design, develop, test, deploy, support, and sustain a product across its lifetime. Moreover, the PMLC WBS contains the standardized work activities necessary to effectively manage and control the project in conformance with management or customer approved constraints of scope, time, budget, quality and resources. Collectively, the three decompositions define most if not all of the work required in a project.

However, the organization may require the integration of other corporate processes within a project, such as acquisition or supplier management. Integration of such work occurs following the same process identified for incorporating project management processes. Simply identify the deliverables associated with each process and integrate them in the WBS. The information necessary to schedule the related work activities should be contained in the process descriptions. As with the systems development and project management processes, hopefully, the organization maintains detailed information on each process within an activity worksheet or some other documentation mechanism used in their process improvement and maturity programs.

I use a five-step process to create the projects WBS:

1. Create a WBS document and start the first line with a multilevel list in the following format:

    1 - First Level

    1.1 - Second Level

    1.1.1 - Third Level

2. Copy the project management process deliverables into the WBS file.
3. Copy required corporate process deliverables into the WBS file.
4. Copy the SDLC Process deliverables into the WBS file.
5. Copy the product deliverables into the WBS file.

We now have all the activities and tasks incorporated into a single WBS. Now we need to create the project schedule. The process can be performed manually in a word processing or spreadsheet tool, but I wouldn't recommend it. Instead, export the data into a project scheduling tool.

The process to create a project schedule starts with constructing the project's network diagram by incorporating the activity dependencies to show the flow of work. This information should be available in the corporate process descriptions or activity worksheets.

At this point, the project's start and stop dates should be added to constrain the project's overall schedule duration. With the dependency information already included for each deliverable, we now need to decide what level of granularity is required to manage our schedule. For example, if each deliverable date falls within the 8/80 rule, we may not need to provide more detail. Simply put in the estimated durations to create each deliverable. That information too should be available in the related process description documents.

Don't forget to add resources to ensure that the schedule takes those constraints into account as it schedules each activity. As a side note, I also recommend that the project manager provide a link to the process descriptions in a note field for each deliverable. This will help ensure that team members performing the work can find the information they need to complete the work.

If the estimated durations for the deliverables do not fit in the 8/80 window, then the PM and SMEs should go into the process descriptions and further decompose the work into smaller increments. At this point it's perfectly acceptable to use the activity names within the schedule, as opposed to the names of associated deliverables. Note that the project scheduler requires information on activity dependencies, estimated durations, and resources to complete the schedule.

SAMPLE PROJECT MANAGEMENT WBS

In this section, I have developed a sample PMLC WBS with the intent of reusing the content via a template. If you like the format, simply copy the template into your project schedule along with the project's product oriented WBS elements.

This template is not meant to imply that these are the only deliverables, sub-deliverables, activities, and tasks that occur across a project management life cycle. This template only serves as a starting point and must be modified to meet the unique requirements of each project. Feel free to create your own template. Taking advantage of reusable templates is almost always better than starting a project WBS and schedule plan from scratch.

PMLC WBS Template

I have documented and maintained a reusable list of common PMLC processes and related templates, incorporating knowledge gained from planning and managing more than 30 enterprise-class IT projects. These practices are documented in the form of a PMLC WBS template, shown in Appendix A. Sample Project Management Life Cycle WBS Template. The template is offered by Command Results (www.commandresults.com) as a downloadable document.

The sample PMLC WBS lists project management deliverable and activities. Note that the type column lists what each item in the decomposition represents as a deliverable, sub-deliverable, or activity. If you prefer, feel free to swap out the term "activity" with "work package." They essentially mean the same thing. No matter which term you use, ideally each item has one or more WBS dictionary references or activity worksheets to describe how to perform the work.

In actual practice, some of the deliverables are performed more than once during the project's duration. An example would be the need to create additional instances of project status reports on a periodic basis. In such cases, the project managers would copy and paste additional instances of each deliverable to be duplicated, as required, within the activity network diagram and Gantt Charts.

Also, it may be necessary to provide more detail for some or all the proposed PM deliverables. In those cases, the project managers only need insert lines to add in additional activities and tasks, at whatever level of detail is necessary to manage the project properly. And, of course, lines can be deleted as well, if they are not necessary to the project.

The sample PMLC WBS template provides a list of the most common project management deliverables and tasks used across a PMLC. However, there is nothing that prevents an assigned program or project manager from adding or deleting WBS elements to fit their organization's or customer's requirements. The point being, plan and schedule at the level of detail you need to manage your projects effectively.

PMLC WBS Template Benefits

The benefits of using predefined PMLC WBS templates include:

- They save time
- They help ensure that the organization consistently uses best practices and Integrating Project Work uses them correctly
- They help project team members discover all the deliverables, activities, and tasks that are included in a project management life cycle
- They help inform discussions with sponsors, customers, and other project team members on the scope of project management activities that are required to support the project.

SYSTEMS DEVELOPMENT WBS

I have also documented and maintained a reusable list of common SDLC-based phases, deliverables, activities, and tasks, incorporating knowledge gained from planning and managing more than 30 enterprise-class IT Projects. These practices are documented in the form of an SDLC WBS Template, shown in Table 18. Sample SDLC WBS Template is offered by Command Results (www.commandresults.com) as a downloadable document.

This sample SDLC WBS template evolved over numerous projects to provide a starting point for IT project planning, and to facilitate discussions with the project team to determine those SDLC WBS elements that make sense for inclusion within an upcoming project. Rather than starting from scratch on each project, this template was reused and modified as required to address the needs of each project.

From one IT project to another, development methods and tools, plus the applications and systems under development, have varied, so this SDLC WBS template is purposely product-and technology-agnostic. Technical details need to be added to the project schedule, usually with assistance from the experts on the project's development team, during the project's planning phase.

The initial SDLC WBS is modified throughout the project as the project team gains a better understanding of the work that's required, and due to the result of change requests that are submitted by the project team members, end users, or customers—as approved by the project's sponsors.

Integrated PMLC - SDLC WBS Template

The SDLC WBS template contained in this section provides a sample WBS covering most IT processes at a fairly high level of detail. Each project team would be expected to remove those SDLC elements that do not apply to the needs of the project and to add additional elements and details as needed.

Notice that the number and types of columns used in this WBS template are different than those used in the previously displayed sample PMLC WBS. In the PMLC WBS template, the column headings were traditional headings found in a project schedule, such as estimated Duration, Resources, Start and Finish Dates, and Dependencies. However, in the SDLC WBS template, we have used columns to capture information by skill and percent utilization of those skills for each defined activity or task. This format is useful for project resource planning and budgeting purposes. Either of the approaches is acceptable, and the project manager may use both WBS formats during the project planning phase: one to help estimate resource and cost estimates, and the other to develop the schedule.

This template already integrates the PMLC WBS, starting at WBS ID Number 1. As simple as the concept is, this is where the PMLC and SDLC integration magic occurs. When we specify the duration, start and stop dates, resources, and dependencies across the WBS, for both the defined PMLC and SDLC activities and tasks, we will have automatically done the work that both helps integrate the life cycles and assists in the development of the integrated project schedule.

WBS Templates Must Be Tailored

Mentioned previously, it's not unusual to have the whole scope of work for a project entirely encompassed within a single SDLC phase. In such cases, users can integrate line items from the PMLC and SDLC WBS templates. (See Appendix A and Appendix B.) Users can also add their unique SDLC work tasks, plus they can delete the SDLC line items that are not required for the authorized scope of work. The tasks identified in the PMLC support the traditional model. However, both the PMLC and SDLC tasks can be modified to organize deliverables and work tasks under iterative development cycles when using an Agile-based SDLC.

Altogether there are 247-line items specified when integrating the PM and SDLC WBS. That may seem like a lot, but I have personally developed and managed an integrated master schedule (KMS) on a program that had 11,000 individual line items. In that program, we incorporated the client's integrated program management WBS template, which had about 450-line items for each project. These program-level WBS templates were applied across each of the 20 projects managed within the program's portfolio, taking us to 9,000-line items in their integrated master schedule. Another 2,000-line items were added to support deployment of the applications across this very large federal health agency. The point is that a very large, complex project and program WBS can be defined much more quickly and simply with reusable WBS templates.

Final comments on scheduling

In a WBS, the identification and listing of phases and deliverables occurs only once for each item. If there are to be multiple development iterations, the resulting schedule quickly grows beyond the 247 deliverable items identified in the template. Also, programs or projects that have multiple and simultaneous activities, such as projects that include applications development, application integration, and process improvement activities, have very large project schedules. But all such requirements can be built from the activities and tasks specified within this integrated PM and SDLC template.

For example, I modified this SDLC WBS template to include Agile iterations over the course of a year plus all schedule milestone reviews and reporting requirements, as implemented by previous clients of mine. That version of my SDLC template has 1,454 schedule line items. The sprints and product releases are prescheduled throughout a year's duration, but the specific product deliverables are not identified. This template allows me to quickly implement a new Agile-based project schedule annually, with the start of each fiscal year, and then fill in the details of specific themes, epics, and user stories as they are identified, prioritized and available for scheduling.

The main point to be made here is that it's much simpler to define a detailed WBS by starting from a standardized and reusable PM/ SDLC WBS template, as opposed to having to start over on every new project. Even on the very largest programs and projects, most of the subprojects are repetitions of the deliverables, activities, and tasks identified in the sample PMLC/SDLC WBS template.

Finally, IT best practices and technology are constantly evolving and moving targets, so this SDLC WBS gets modified and updated a lot. Sometimes changes are made at the start of every new project. With this caution in mind, feel free to use this template and evolve it during your career to support your own SDLC planning needs.

CHAPTER SUMMARY

In this chapter, I have covered a very broad topic on how to integrate project management, systems development, and other corporate processes within a WBS. Key to understanding project integration is knowing how to decompose products, processes, and work as a series of deliverables. A WBS lists the things that are delivered by the project or development team but should not provide a listing of how the development and project work is performed. The information governing processes, activities, and tasks is maintained in the organization's process documentation.

Ultimately, the deliverables are consolidated within a single WBS. This is the first step toward creating the project's schedule. Project schedules are formed by adding dependency relationships, tasks durations, and resource assignments to the deliverables identified in the WBS. If the work associated with a deliverable exceeds the 8/80 hours rule, the project scheduler and SMEs should dive into the process documentation to list the activities and tasks of the processes associated with each deliverable. Again, task dependency relationships, tasks durations, and resource assignments are required to define the project's detailed schedule.

In situations where multiple product or project teams are working in parallel on different components of the overall solution, the schedules must be integrated. The integration can start in the WBS, or the integration can occur in a schedule. A schedule that includes information from multiple product and project teams is called an integrated master schedule.

Not discussed, but also possible, is to align the inputs and outputs of multiple schedules manually. Given the size and complexity of some programs and projects, that is not my preference. I've had experiences where other project managers hid schedule delays, causing huge problems when they were unable to provide their deliverables at the program level on a timely basis. The issue with hiding their problems is that they could not get the management attention they required to address their problems.

We learned in this chapter that it's rare for a single project to encompass an entire product, system, or solution life cycle. Rather, it's more practical to break up product life cycle work into smaller chunks to manage risks and provide more opportunities for reviews and adjustments, and more time and to evaluate investment decisions on an ongoing basis. Also, product-related enhancements often extend out over multiple fiscal years, requiring new executive reviews, budgets, and approvals.

Finally, this chapter presented a project management WBS and an integrated project management and systems development WBS, which can support either traditional and Agile practices with tailoring. These templates can be downloaded to serve as a starting point for IT project planning, and to facilitate discussions with a project team to determine those SDLC WBS elements that make sense for inclusion in an upcoming project. The integrated WBS also helps management, customers, and development teams understand and agree on the approach and scope of work and project management attention required to support their project.

Conclusion

This ends the first book in the series. This book is foundational, providing the basic information on systems development that IT project members should have to understand the nomenclature and the issues involved in developing complex software and systems. This may be the end of your journey of knowledge with me. However, there is a lot more to learn, if you are interested.

In the next book, titled *Frameworks for Agility*, I'll dive into more detail on both engineering and Agile processes. Both sets of processes are required to build the right set of products, efficiently and collaboratively. The book describes in detail the use of Scrum processes as the reference model for implementing Agile-based practices. Moreover, my Frameworks for Agility book provides a sample hybrid SDLC framework and model to maintain both sets of processes and related information artifacts, tools and templates.

My third book, titled Tools and Templates, is for those readers who want more information on the tools and templates used in systems development and project management processes. Detailed descriptions of the templates are provided in context with how and when they are used, along with descriptions of the data elements typically associated with each template.

My fourth book is titled DevOps. This is the most technical of the four books. The book covers the topics of CI, CD, DevOps, Kanban, Scrumban, JIT, and lean practices in much greater detail than this foundational book. Moreover, the book also gets into the technical details of software development, and some of the tools available today, that can help fulfill the goal of implementing SDLC process integration and automation capabilities promoted under CI, CD, and DevOps concepts. The reason this is the fourth book in the SDLC series is that we must thoroughly understand the purpose of SDLC processes and how they work before we can hope to make them efficient through process integration and automation.

Appendix A. Sample Project Management Life Cycle WBS Template

WBS ID No	Name	Type	Duration	Resources	Start	Finish	Dependencies
0	Project WBS						
1	Project Management Deliverables	Deliverable					
1.1	Project Charter	Deliverable					
1.1.1	Define project's purpose	Task					
1.1.2	Define business need	Task					
1.1.3	Define ROI and cost-benefit justifications	Task					
1.1.4	Description of deliverables	Task					
1.1.5	Define measures of success	Task					
1.1.6	List critical stakeholders	Task					
1.1.7	List participating org. departments	Task					
1.1.8	Define project constraints	Task					
1.1.9	List assumptions	Task					
1.1.10	List important project milestones	Task					
1.1.11	Document budget summary	Task					
1.1.12	Name project manager & authorities	Task					
1.1.13	Statement of work (SOW)	Task					
1.2	Project Management Plan	Deliverable					
1.2.1	Project scope statement	Sub-deliverable					
1.2.2	Resource sub plan	Sub-deliverable					
1.2.3	Communications sub plan	Sub-deliverable					
1.2.4	Risk management sub plan	Sub-deliverable					
1.2.5	Change control sub plan	Sub-deliverable					
1.2.6	Quality assurance sub plan	Sub-deliverable					
1.2.7	Procurement sub plan	Sub-deliverable					
1.2.8	Budget sub plan	Sub-deliverable					
1.2.9	Project team members, roles & responsibilities	Sub-deliverable					
1.2.10	Project schedule	Sub-deliverable					
1.3	Project Status Meetings	Deliverable					
1.3.1	Project kickoff meeting	Deliverable					
1.3.2	Create weekly status report	Task					
1.3.3	Facilitate weekly status report meetings	Task					
1.3.4	Facilitate monthly client review meetings	Task					
1.4	Project Administration	Deliverable					
1.4.1	Onboard project team members	Task					
1.4.2	Take meeting notes	Task					
1.4.3	Obtain & record budget updates	Task					
1.4.4	Establish project performance standards	Task					
1.4.5	Establish project reporting standards	Task					
1.4.6	Establish project naming conventions	Task					
1.5	Project Risk Management	Deliverable					

1.5.1	Develop risk identification form	Task
1.5.2	Facilitate risk/issues review meetings	Task
1.5.3	Update risk/issues Log	Task
1.6	**Project Change Management**	**Deliverable**
1.6.1	Develop change request form	Task
1.6.2	Facilitate change request meetings	Task
1.6.3	Update change request log	Task
1.7	**Project Lessons Learned**	**Deliverable**
1.7.1	Develop lessons learned log	Task
1.7.2	Update lessons learned	Task
1.8	**Project Actions Items**	**Deliverable**
1.8.1	Develop action items log	Task
1.8.2	Record action items	Task
1.8.3	Close action items	Task
1.9	**Phase Gate/Milestone Reviews**	**Deliverable**
1.9.1	Review exit criteria from the previous phase	Task
1.9.2	Review entry criteria for next phase	Task
1.9.3	Ensure phase deliverables are complete	Task
1.9.4	Prepare review presentation materials	Task
1.9.5	Facilitate PG/milestone review meetings	Task
1.9.6	Closeout PG/milestone review action items	Task
1.10.	**Project Close**	**Deliverable**
1.10.1	Submit certification of compliance, completion & approval form	Task
	Obtain final sign-off on CC&A form	Task
1.10.2	Capture remaining lessons learned	Task
1.10.3	Close-out lessons learned log	Task
1.10.4	Release project team members	Task
1.10.5	Close-out project documentation	Task

Appendix B. Sample SDLC WBS Template

WBS No	Task Name	Estimated Duration (%)	PM (%)	BA (%)	Architect (%)	GUI Dev (%)	Tech Cons (%)	SW Dev (%)	DB DEV (%)	QA/ Test (%)	Total Hrs.
0	Project WBS	5	10%	100%		33%	33%				8.8
1	Project Management										0
1.1.	Project management plan										0
1.1.1.	Develop project charter										0
1.1.2.	Define scope										0
1.1.3.	Define resource plan										0
1.1.4.	Develop communications plan										0
1.1.5.	Develop risk plan										0
1.1.6.	Develop change control plan										0
1.1.7.	Develop quality plan										0
1.1.8.	Develop purchase plan										0
1.1.9.	Develop cost plan										0
1.1.10.	Develop organization plan										0
1.1.11.	Develop project schedule										0
1.2.	Project meetings										0
1.2.1.	Conduct kickoff meeting										0
1.2.2.	Weekly status meetings										0
1.2.3.	Monthly tactical meetings										0
1.2.4.	Project closing meeting										0
1.3.	Administration										0
1.3.1.	Document performance standards										0
1.3.2.	Document reporting standards										0
1.3.3.	Document naming conventions										0
1.3.4.	Document housekeeping standards										0
	Sum Total =>	5	10%	100%	33%	0%	33%	0%	0%	0%	8.8
2	Business Requirements Document										0
2.1.	Cross-organizational steering committee										0
2.1.1.	Identify key stakeholders										0
2.1.2.	Identify steering committee members										0
2.1.3.	Form steering committee										0
2.1.4.	Conduct kickoff meeting										0
2.1.5.	Conduct advisory meetings										0
2.2.	High-level business requirements										0
2.2.1.	Strategic requirements										0
2.2.1.1.	Conduct interviews with executive management										0
2.2.1.2.	Identify relevant strategic requirements										0
2.2.1.3.	Document strategic requirements										0
2.2.2.	Stakeholder requirements										0
2.2.2.1.	Identify critical stakeholders										0

2.2.2.2.	Conduct interviews with identified stakeholders										0
2.2.2.3.	Identify relevant stakeholder requirements										0
2.2.2.4.	Document stakeholder requirements										0
2.2.3.	Customer requirements										0
2.2.3.1.	Conduct interviews with representative customers										0
2.2.3.2.	Identify relevant customer requirements										0
2.2.3.3.	Document customer requirements										0
2.3.	As-is process										0
2.3.1.	Meet with process owners										0
2.3.2.	Create summary level "as-is" process map										0
2.3.3.	Document current processes and practices										0
2.3.4.	Document critical business challenges										0
2.3.5.	Document management & process practitioner challenges										0
2.4.	To-be process requirements										0
2.4.1.	Identify "Capabilities" not currently available but required by the client										0
2.4.2.	Research comparable "best practice"										0
2.4.3.	Document best practices										0
2.4.4.	Create summary level "to-be" process map										0
2.4.5.	Develop detailed process & workflow requirements document										0
2.5.	Perform gap analysis										0
2.5.1.	Determine gaps between the current "as-is" and the proposed "to-be" states										0
2.5.1.1.	Identify people skills and resources gaps										0
2.5.1.2.	Identify process change requirements										0
2.5.1.3.	Identify technology gaps										0
2.5.1.4.	Identify information gaps										0
2.5.2.	Final review and acceptance										0
2.6.	Develop as-is/ to-be transition plan										0
2.6.1.	Research transition approaches										0
2.6.2.	Create draft transition plan										0
2.6.3.	Review draft transition plan										0
2.6.4.	Update draft transition plan										0
2.6.5.	Review final transition plan										0
2.6.6.	Final review and acceptance										0
	Sum Total =>	0	0%	0%	0%	0%	0%	0%	0%	0%	0
3	Solution System Design										0
3.1.	Use cases										0
3.1.1.	Conduct end-user interviews										0
3.1.2.	Develop uses cases describing system interaction requirements										0
3.1.3.	Review uses cases with end-users and management										0
3.1.4.	Review use cases with steering committee r epresentatives										0
3.1.5.	Update use cases, as necessary										0
3.1.6.	Final review and acceptance										0
3.2.	Screen prototypes and mockups										0
3.2.1.	Analyze functional requirements										0
3.2.1.1.	Analyze workflow requirements										0
3.2.1.2.	Analyze screen display & data entry requirements										0
3.2.1.3.	Analyze data requirements										0

3.2.2.	Develop initial mockup screens										0
3.2.3.	Show prototype/ mockup screens to end-users										0
3.2.3.1.	Document requested enhancements										0
3.2.4.	Show prototype/ mockup screens to management										0
3.2.4.1.	Document requested enhancements										0
3.2.5.	Show prototype/ mockup screens to steering committee members										0
3.2.5.1.	Document requested enhancements										0
3.2.6.	Make final changes to prototype/ mockup screens										0
3.3.	System requirements documentation										0
3.3.1.	Create draft system requirements										0
3.3.2.	Review draft system requirements										0
3.3.3.	Update draft system requirements										0
3.3.4.	Review final system requirements										0
3.3.5.	Final review and acceptance										0
3.4.	User documentation										0
3.4.1.	Create draft user documentation										0
3.4.2.	Review draft user documentation										0
3.4.3.	Update draft user documentation										0
3.4.4.	Review final user documentation										0
3.4.5.	Final review and acceptance										0
3.5.	Training program materials										0
3.5.1.	Establish initial training requirements & goals										0
3.5.2.	Review & approve training requirements										0
3.5.3.	Create initial training materials										0
3.5.4.	Review and approve training materials										0
3.5.5.	Conduct trial course delivery										0
3.5.6.	Update and finalize training materials										0
3.5.7.	Final review and acceptance										0
3.6.	Hardware requirements										0
3.6.1.	Create draft hardware requirements										0
3.6.2.	Review draft hardware requirements										0
3.6.3.	Hardware requirements approved										0
3.7.	Implementation and support requirements										0
3.7.1.	Create draft implementation and support requirements										0
3.7.2.	Review draft implementation and support requirements										0
3.7.3.	Implementation and support requirements approved										0
	Sum Total =>	0	0%	0%	0%	0%	0%	0%	0%	0%	0
4	COTS Product Assessments										0
4.1.	DAR assessment										0
4.1.1.	Establish decision analysis and resolution (DAR) evaluation team										0
4.1.2.	Formulate decision objectives and scope										0
4.1.3.	Decide COTS selection criteria and weighting										0
4.1.4.	Identify COTS product alternatives										0
4.1.5.	Select decision-making method										0
4.1.6.	Rate criteria for each alternative										0
4.1.7.	Develop composite scores for each COTS product										0
4.1.8.	Select and communicate best fit COTS alternative										0
	Sum Total =>	0	0%	0%	0%	0%	0%	0%	0%	0%	0
5	Detail Software & Hardware Configurations										0

5.1.	Create initial software & hardware configuration design										0
5.2.	Review initial software & hardware c onfiguration design										0
5.3.	Update initial software & hardware configuration design										0
5.4.	Review final software & hardware configuration design										0
5.5.	Software design & hardware configuration design approved										0
	Sum Total =>	0	0%	0%	0%	0%	0%	0%	0%	0%	0
6	System & Hardware Implementation										0
6.1.	Implementation plan										0
6.1.1.	Create initial software & hardware implementation plan										0
6.1.2.	Review initial software & hardware implementation plan										0
6.1.3.	Update initial software & hardware implementation plan										0
6.1.4.	Review final software & hardware implementation plan										0
6.1.5.	Software design & hardware implementation plan approved										0
6.2.	User, administration, and training documentation										0
6.2.1.	Customize user documentation										0
6.2.2.	Customize administration documentation										0
6.2.3.	Customize training program materials										0
6.3.	Install hardware & software										0
6.3.1.	East coast data center location										0
6.3.2.	Regional offices										0
6.4.	Support hardware & software										0
	Sum Total =>	0	0%	0%	0%	0%	0%	0%	0%	0%	0
7	Custom Software Development										0
7.1.	Review function and technical specifications										0
7.2.	Develop mock user interface and navigational screens and menus (See section 3.2)										0
7.3.	Review with prospective end users										0
7.4.	Incorporate agreed user desired changes										0
7.5.	Final review of mock presentation and navigational screens and menus										0
7.6.	First iteration application build										0
7.6.1.	Build data model										0
7.6.2.	Build DBMS										0
7.6.3.	Build logic constructions										0
7.6.4.	Integrate presentation, navigation and DBMS components										0
7.6.5.	Test										0
7.6.6.	Review with prospective end users										0
7.7.	Second iteration application build										0
7.7.1.	Build data model										0
7.7.2.	Build DBMS										0
7.7.3.	Build logic constructions										0
7.7.4.	Integrate presentation, navigation and DBMS components										0

7.7.5.	Test										0
7.7.6.	Review with prospective end users										0
7.8.	Third iteration application build										0
7.8.1.	Build data model										0
7.8.2.	Build DBMS										0
7.8.3.	Build logic constructions										0
7.8.4.	Integrate presentation, navigation and DBMS components										0
7.8.5.	Test										0
7.8.6.	Review with prospective end users										0
	Sum Total =>	0	0%	0%	0%	0%	0%	0%	0%	0%	0
8	Integration & Testing										0
8.1.	Software test planning										0
8.1.1.	Software test plan										0
8.1.2.	Software test cases										0
8.1.3.	Software test results										0
8.1.4.	Acceptance test plan										0
8.1.5.	Acceptance test cases										0
8.1.6.	Acceptance test results										0
8.2.	Hardware test planning										0
8.2.1.	System test plan										0
8.2.2.	System test cases										0
8.2.3.	System test results										0
8.2.4.	Acceptance test plan										0
8.2.5.	Acceptance test cases										0
8.2.6.	Acceptance test results										0
8.3.	Solution testing										0
8.3.1.	Unit test										0
8.3.2.	Component integration testing										0
8.3.3.	Systems integration testing										0
8.3.4.	Quality assurance testing										0
8.3.5.	Performance testing										0
8.3.6.	Regression testing										0
8.3.7.	Production/systems acceptance testing										0
	Sum Total =>	0	0%	0%	0%	0%	0%	0%	0%	0%	0
9	Deployment										0
9.1.	Deployed software										0
9.1.1.	Software user documentation										0
9.1.1.1.	Software user guides										0
9.1.1.2.	Software systems administration guides										0
9.1.1.3.	Software job aids										0
9.1.2.	Software training program materials										0
9.1.2.1	Create training outline										0
9.1.2.2	Create training storyboards										0
9.1.2.3	Create training artifacts (Web, job aids, Wikis, etc.)										0
9.1.2.4	End-user, customer, stakeholder reviews										0
9.1.2.5	End-user, customer, stakeholder signoff										0
9.1.2.6	Upload/ deploy training materials/artifacts										0
9.1.2.7	Software deployment communications										0
9.2.	Deployed hardware										0
9.2.1.	Hardware user documentation										0
9.2.2.	Hardware training program materials										0
9.2.2.1	Create training outline										0
9.2.2.2	Create training storyboards										0

Appendix B

9.2.2.3	Create training artifacts (Web, job aids, Wikis, etc.)										0
9.2.2.4	End-user, customer, stakeholder reviews										0
9.2.2.5	End-user, customer, stakeholder signoff										0
9.2.2.6	Upload/ deploy training materials/ artifacts										0
9.2.3.7	Hardware deployment communications										0
9.3.	Deployed processes										0
9.3.1.	Process user documentation										0
9.3.2.	Process training program materials										0
9.3.2.1	Create training outline										0
9.3.2.2	Create training storyboards										0
9.3.2.3	Create training artifacts (Web, job aids, Wiki's, etc.)										0
9.3.2.4	End-user, customer, stakeholder reviews										0
9.3.2.5	End-user, customer, stakeholder signoff										0
9.3.2.6	Upload/deploy training materials/artifacts										0
9.3.2.7	Process deployment communications										0
	Sum Total =>	0	0%	0%	0%	0%	0%	0%	0%	0%	0
	Grand Total =>	5	10%	100%		33%	0%	33%	0%	0%	0%

List of Acronyms

AD	Agile data
AM	agile modeling
AMDD	Agile model-driven development
APM	alternative payment model
AUP	Agile unified process
BDD	behavior-driven development
BI	business intelligence
BPI	Business process improvement
BPR	business process re-engineering
BRD	business requirements document
BSI	British Standards Institution
CAD	computer-aided design
CASE	computer-aided software engineering
CD	continuous delivery
CD	continuous deployment
CD/D	continuous delivery and deployment
CI	configuration item
CI	continuous integration
CM	configuration management
CMM	capability maturity model
CMMI	capability maturity model integration
COBIT	control objectives for information and related technology
COM	computer operation manual
CONOPS	concept of operations
COTS	commercial-off-the-shelf
CPM	computer programming manual
CSM	Certified Scrum Master
CSPO	Certified Scrum Product Owner
CT	continuous testing
DA	disciplined Agile
DAC	Disciplined Agile Consortium
DAD	disciplined Agile delivery
DAE	disciplined Agile enterprise
DAR	decision analysis and resolution
DBDD	database design description
DBMS	database management system
DCF	discounted cash flows
DevOps	Development - operations
DFSS	design for six sigma
DID	data item description
DMADV	Define, measure, analyze, design, verify
DMAIC	Define, measure, analyze, improve, control

List of Acronyms

DMM	data maturity management
DoD	Department of Defense
DSD	design specification document
DSDM	Dynamic Systems Development Model
DSL	domain-specific language
ECM	Enterprise Change Model
ETL	Extract, transform, load
EPF	Eclipse process framework
EPPM	enterprise project portfolio management
EUP	enterprise unified process
FDASIA	Food and Drug Administration Safety and Innovation Act
FDD	feature-driven development
FF	finish-to-finish
FFP	firm fixed price
FS	finish-to-start
FSD	functional specification document
FSM	firmware support manual
GEP	good engineering practice
GIS	geographic information systems
GUI	graphical user interface
HIPAA	Health Insurance Portability and Accountability Act
HIT	Health information technology
HITECH	Health Information Technology for Economic and Clinical Health
ICASE	integrated computer aided software engineering
IDD	interface design description
IDEs	integrated development environments
IDEF	Integrated Definition
IEEE	Institute of Electrical and Electronics Engineers
IID	iterative and incremental development
IMS	integrated master schedule
INVEST	Independent, negotiable, valuable, estimate-able, small, testable
IoT	Internet of things
IRS	interface requirements specification
ISMS	information security management system
ISO	International Organization for Standardization
ISO/IEC	International Organization for Standardization/International Electrotechnical Commission
IT	information technology
ITG	Information technology governance
ITIL	formerly, information technology infrastructure library
ITSM	IT service management
IV&V	independent verification and validation
JAD	joint application development
JIT	just in time
MACRA	Medicare Access and CHIP Reauthorization Act of 2015
MDD	model-driven development
MDDAP	Medical Device Discovery Appraisal Program
MIL	Military, as in military standard
MIPS	merit-based incentive payment system
MOF	managed object format
O&M	Operations & Maintenance
OCD	operational concept description
OCM	organization change management
OMG	object management group
OMT	object-modeling technique

List of Acronyms

OO	object-oriented
OOD	object-oriented development
OOP	object-oriented programming
OOSE	object-oriented software engineering
Open UP	Open Unified Process
PBS	Product Breakdown Structure
PDCA	plan–do–check–act, or plan–do–check–adjust
PDDQ	patient demographic data quality
PDSA	plan–do–study–act
PERT	Program Evaluation Review Technique
PIT	process improvement team
PMI	Project Management Institute
PMLC	project management life cycle
PMO	program management office
PPM	project portfolio management
PSP	personal software process
QA	quality assurance
RACI	Responsible, accountable, consulted, informed
RASCI	Responsible, accountable, supporting, consulted, informed
RAM	responsibility assignment matrix
RAD	Rapid Application Development
RFI	request for information
RFP	request for proposal
RFQ	request for quote
RIF	risk identification form
RMC	rational method composer
ROI	Return on Investment
RTM	requirements traceability matrix
RUP	Rational unified process
SA	structured analysis
SAD	software/system architecture document
SAFe	Scaled agile framework
SAGE	semi-automatic ground environment
SCAMPI	Standard CMM-I Appraisal Method for Process Improvement
SCOM	software center operator manual
SD	structured design
SDC	Systems Development Corporation
SDD	software design description
SDLC	systems/ systems/ solutions development life cycles
SDP	software development plan
SEI	Software Engineering Institute
SF	Start-to-Finish
SIOM	software input/output manual
SIP	software installation plan
SMEs	subject matter experts
SOW	statement of work
SOX	Sarbanes-Oxley Act of 2002
SPS	software product specification
SQA	Software Quality Assurance
SQC	statistical quality control
SROI	social return on investment
SRS	software Requirements Specification
SS S	tart-to-Start
SSADM	structured systems analysis and design method
SSDD	system/subsystem design description

List of Acronyms

SSS	system/subsystem specification
STD	software test description
STP	software test plan
STR	software test report
STP	software transition plan
SUM	software user manual
SVD	software version description
TDD	test-drive development
TOGAF	the open group architecture framework
TRM	technical reference model
TSP	team software process
UAT	user acceptance testing
UCD	user-centered design
UML	unified modeling language
WBS	work breakdown structure
WIP	work in process
XP	extreme programming

Index

A
Agile, 3–4, 8–10, 12, 15, 17–19, 21–25, 27–29, 33–35, 37–41, 43–45, 50, 53, 55–57, 60, 63, 67–69, 73, 75–84, 91–92, 94–95, 97–99, 101, 103–107, 109–111, 116–117, 119–123, 127–132, 145–146, 149–150, 155, 157–158, 161–167, 170, 173, 175–178, 188, 193, 200, 211, 216, 218–224, 226, 230–231, 236–237, 241–244, 247, 249, 251–252, 254, 256–264
Agile model-driven development, 95, 101, 247
agile modeling, 104, 247, 251, 256, 262
Agile unified process, 101, 247, 251, 261
alternative payment model, 247, 251

B
behavior-driven development, 84, 247, 251
British Standards Institution, 154, 247, 251
business intelligence, 183–184, 247, 251
Business process improvement, 16, 247, 251
business process re-engineering, 16, 247
business requirements document, 49, 160, 221, 247

C
capability maturity model, 33, 108, 247, 251
capability maturity model integration, 33, 247
Certified Scrum Master, 23, 247, 251
Certified Scrum Product Owner, 23, 247, 251
commercial-off-the-shelf, 43, 81, 247, 251
commercial-off-the-shelf (COTS), 43, 81
computer operation manual, 174, 247, 251
computer programming manual, 174, 247, 251
computer-aided design, 129, 247, 251
computer-aided software engineering (CASE), 129, 247
concept of operations (CONOPS), 49, 247, 251
configuration item (CI), 27, 247, 251
configuration management (CM), 26, 117, 123, 247, 251
continuous delivery (CD), 11, 76, 83, 89–90, 92, 247, 251
continuous delivery and deployment, 11, 247
continuous deployment (CD), 90–92, 247, 251
continuous integration (CI), 11, 29, 64, 83, 87–88,
91, 98, 159, 177–178, 222, 247, 251
continuous testing, 11, 83, 89–91, 247, 251
control objectives for information and related technology (COBIT), 154, 247

D
data item description, 247, 251, 264
data maturity management, 109, 247, 251
database design description, 174, 247, 251
database management system, 183, 247, 251
decision analysis and resolution, 247
Define, 16, 18, 22, 26, 28, 33–34, 40, 48–49, 53, 65–66, 69, 71, 78, 94, 96, 100, 143, 145, 154, 159, 164, 173, 176, 182–184, 192, 194–195, 199–204, 206–211, 213, 217–219, 224–230, 232–233, 238, 242, 247, 251
Department of Defense (DOD), 47–48, 101, 108, 123,
172, 247, 251, 264
Design, 6, 22–24, 26–27, 29, 31–32, 38, 40–44, 46–47, 49–52, 54–62, 66–72, 74–75, 77, 82, 87, 95–97, 100, 112–113, 115, 118–131, 140, 142–148, 156, 160–161, 163–166, 169–170, 172–175, 187, 191, 194–195, 198, 201, 206–207, 213, 220–224, 227, 230, 238, 247–252, 254–256, 260, 263
design for six sigma, 97, 247, 251
design specification document, 115, 247, 251
Development - operations, 247, 251
disciplined Agile, 99, 103–104, 120–121, 130, 132, 247, 252, 256–257, 262
Disciplined Agile Consortium, 103, 247, 252, 256, 262
disciplined Agile delivery, 103–104, 247, 252, 256–257, 262
disciplined Agile enterprise, 104, 247, 252
discounted cash flows, 139, 247, 252
DMADV, 3, 96–97, 247, 252
DMAIC, 3, 96–97, 247, 252
domain-specific language, 84, 94, 248, 252
Dynamic Systems Development Model, 57, 104, 247

E
Eclipse process framework, 106, 110, 248, 252
Enterprise Change Model, 2–3, 30, 180, 190–191, 193, 248, 252
enterprise project portfolio management, 141, 248
enterprise unified process, 101, 248, 252, 262
Extract, 74, 183, 248, 252

extreme programming (XP), 38, 99, 105, 250, 252, 256, 260, 262

F
feature-driven development, 95, 248, 252
finish-to-finish, 4, 215–216, 248, 252
finish-to-start, 3, 215, 248, 252
firm fixed price, 146, 248, 252
firmware support manual, 174, 248, 252
Food and Drug Administration Safety and Innovation Act, 153, 248
functional specification document, 115, 248

G
geographic information systems (GIS), 154, 248
good engineering practice, 119, 248, 252
graphical user interface (GUI), 82, 248, 252

H
Health information technology, 153, 248, 252
Health Information Technology for Economic and Clinical Health, 153, 248
Health Insurance Portability and Accountability Act, 153, 248

I
IID, 37–38, 43–45, 50, 55, 57, 59–73, 77, 81, 95, 105, 112, 121, 123, 126–129, 161, 163, 176–177, 179, 203, 221–223, 230, 248, 252
Incremental, 3–4, 23–24, 33, 37, 43, 56, 60–62, 64–65, 67–68, 72, 75, 81, 86, 112, 129, 164, 170, 172, 188, 222, 226, 248, 252–253, 256, 259, 264
Independent, 29, 41, 87, 102, 107–108, 150, 193, 198, 227, 236, 248, 252, 262
independent verification and validation (IV&V), 248
information security management system, 212, 248
information technology (IT), 11–12, 19, 28, 104, 114, 136, 150–154, 159, 185, 248, 252, 263
Information technology governance, 248
Institute of Electrical and Electronics Engineers (IEEE), 248
integrated computer aided software engineering, 100, 248
Integrated computer aided software engineering (ICASE), 100
Integrated Definition, 248, 252
integrated development environments, 57, 69, 82, 248
integrated master schedule (IMS), 219, 241–242, 248, 252
interface design description, 174, 248, 252
interface requirements specification, 174, 248
International Organization for Standardization, 154, 212, 248
International Organization for Standardization/International Electrotechnical Commission, 154, 248
Internet of things (IoT), 158, 248, 253
IT service management, 152, 154, 212, 248, 253
Iterative, 3–4, 17, 24, 27, 37, 39, 43, 48, 50–51, 53, 55–56, 59–68, 70–71, 75, 81, 102–103, 105–106, 110, 120, 122, 125–126, 128, 130–131, 142, 146–147, 161–164, 166, 169–173, 175, 177–178, 193, 216, 222, 224, 241, 248, 253, 259, 264
iterative and incremental development (IID), 3–4, 37, 43, 65, 67–68, 81, 164, 170, 248, 264

J
joint application development (JAD), 39, 248, 253
just in time (JIT), 39, 83, 92, 97, 248, 253

L
lean, 10–11, 17, 80, 83–85, 92–95, 97, 102, 104, 107, 110–111, 136, 155, 159, 162–163, 178, 189, 230, 244, 253, 257–258, 260–261, 263

M
managed object format, 154, 248, 253
Medical Device Discovery Appraisal Program, 109, 248
Medicare Access and CHIP Reauthorization Act of 2015, 153, 248
merit-based incentive payment system, 153, 248
Methods, 9–10, 12–13, 16, 18, 20–23, 25, 29, 32, 35, 39–45, 47, 50, 57, 59, 62, 67–69, 72, 77–78, 80–81, 83, 88, 90, 95, 100–101, 103–104, 110, 120–123, 129–130, 132, 136–137, 140, 143, 149–150, 155–156, 158, 160, 163–165, 168–169, 175, 178–179, 183–184, 189, 196, 201, 203–204, 209, 211–213, 218–219, 222, 226, 229, 240, 253, 261
Military, 47, 123, 172, 248, 253, 264
model-driven development, 43, 57, 94–95, 101, 247–248, 253

O
object management group (OMG), 110, 129, 212, 248, 253, 263–264
object-modeling technique, 249, 253
object-oriented, 39, 70, 110, 112, 128, 249, 253, 256–257, 260
object-oriented development, 70, 249, 253
object-oriented programming, 112, 128, 249, 253
object-oriented software engineering, 110, 249, 257

Open Unified Process, 106, 249, 253
operational concept description, 174, 248
Operations & Maintenance, 160, 248, 253
organization change management, 185, 248

P
patient demographic data quality, 109, 249
personal software process, 108, 249, 253
plan-do-study-act (PDSA), 65, 249, 253
Process, 2–4, 8–24, 26–35, 38–39, 41–42, 44–45, 47–52, 54–55, 57–58, 60–73, 76, 80–89, 91–97, 101–104, 106, 108–112, 114–118, 120, 123–124, 126–129, 132, 136, 138, 146–147, 149–151, 154–155, 158, 160–163, 165–167, 170–173, 176–180, 187–189, 193–200, 202–211, 213–214, 217–218, 221–222, 224–226, 228–230, 232–233, 236, 238–239, 241–242, 244, 247–256, 258, 261–263
process improvement team, 196, 249, 253
Product Breakdown Structure, 4, 235–236, 238, 249, 253
Program, 3, 9, 11–15, 26, 28, 32–33, 35, 40–41, 45–46, 49, 67, 92, 100, 103, 108–109, 118, 126, 130, 134–136, 138–142, 148, 153, 155, 171–173, 175, 178, 186, 195, 200, 203, 206–207, 210, 212–213, 216–217, 219–220, 223, 240–242, 248–249, 253
Program Evaluation Review Technique, 249
program management office, 249, 253
Project Management Institute (PMI), 78, 177, 249, 253, 257–258, 264
project management life cycle, 2, 29–30, 177, 218, 229, 232, 239–240, 245, 249, 253
project portfolio management, 140–141, 248–249, 253
project portfolio management (PPM), 141

Q
quality assurance (QA), 33, 49, 160, 233, 249, 253–254

R
RAD, 4, 37, 43, 45, 55, 57–58, 60, 68–73, 81, 95, 101, 105, 112, 127–129, 249, 253, 257
Rapid Application Development, 68, 81, 249, 253, 257, 259
rational method composer, 110, 249, 253
Rational unified process (RUP), 39, 249, 253
request for information, 142, 249, 253
request for proposal, 142, 249, 253
request for quote, 145, 249, 253
Requirements, 18, 22–27, 29, 31–32, 35, 38–47, 49–78, 82, 84–85, 87, 90–91, 94–95, 97, 100, 105–107, 110, 112–115, 120–132, 135–136, 138, 140, 142–153, 155–156, 160–161, 163–166, 169–171, 173–179, 181–183, 187, 191–192, 194–197, 200–202, 206–208, 210–211, 213–214, 217, 220–224, 226–228, 230, 232–233, 236–237, 239–242, 247–254, 256
requirements traceability matrix, 125, 249
responsibility assignment matrix (RTM, 206, 210, 249
Responsible, 21, 32, 73, 111, 132, 139, 148, 206, 210, 232, 249, 253–254
Return on Investment (ROI), 60, 139, 249–250, 254
risk identification form, 233, 249, 254

S
Sarbanes-Oxley Act of 2002, 22, 249, 254
Scaled-Agile Framework (SAFe), 67, 249, 254, 257
SDLC, 2–4, 6–13, 15–37, 39–40, 42, 44–45, 47–50, 53, 56, 60, 64, 69, 72–73, 75, 77–78, 80–81, 83–86, 89–91, 94, 98–105, 107–108, 111–112, 114, 116–117, 119–124, 127, 130–132, 134–136, 140, 142–143, 145–147, 149–152, 154–156, 158–171, 173, 175–179, 181, 187–189, 191, 193, 198–202, 204–214, 216–225, 227–230, 232–233, 236–238, 240–244, 246, 249, 254, 256–257, 259
semi-automatic ground environment, 178, 249
social return on investment, 139, 250, 254
software center operator manual, 174, 249
software design description, 174, 249, 254
Software Development, 8, 12, 16, 18, 24–25, 35, 37–40, 42, 44–52, 55–57, 65–66, 68–69, 72–76, 78, 81–82, 84–85, 89, 92–95, 97, 100–101, 104–108, 110–112, 114, 116, 118, 120, 122, 128–129, 132, 155, 158–160, 169, 171–175, 178, 226, 244, 249, 254, 256–257, 259–264
software development plan, 174, 249, 254
Software Engineering Institute, 108, 249, 254, 262
software input/output manual, 249, 254
software installation plan, 174, 249, 254
software product specification, 174, 249, 254
Software Quality Assurance (SQA), 249, 254
software Requirements Specification, 174, 250
software Requirements Specification (SRS), 174
software test description, 174, 250, 254
software test plan, 174, 250, 254
software test report, 174, 250, 254
software transition plan, 174, 250, 254
software user manual, 250, 254
software version description, 174, 250, 254
software/system architecture document, 115, 249
Standard CMM-I Appraisal Method for Process Improvement, 249

Start-to-Finish, 3, 215, 249, 254
start-to-Start, 3, 215–216, 254
statement of work, 249, 254
statistical quality control, 95, 250, 254
structured analysis, 47, 249, 254
structured design, 47, 249, 254
structured systems analysis and design method (SSADM), 250
subject matter experts (SME), 26, 32, 249, 254
system/subsystem design description, 174, 250
system/subsystem specification, 174, 250
Systems Development Corporation, 172, 249
systems/ systems/ solutions development life cycles, 249

T
Target market, 31, 91, 135–136, 182, 188, 190–192, 254
team software process, 108, 250, 254, 256, 263
technical reference model, 120, 250, 254
test-driven development (TTD), 250, 254
the open group architecture framework (TOGAF), 154, 250

U
unified modeling language (UML), 94, 250, 254
user acceptance testing, 41, 49, 89, 114, 125, 146, 194, 222, 250, 255
user-centered design, 250, 255

W
work breakdown structure, 33, 250, 255
work in process (WIP), 39, 93–94, 250, 255

Suggested Reading

Ambler, S. W. (2002). Agile Modeling, Effective practices for eXtreme Programming and the Unified Process. New York: Wiley Computer Publishing.

Ambler, S. W., & Lines, M. (2012). Disciplined Agile Delivery: A practitioners guide Agile Software Delivery in the Enterprise. Boston: Pearson Education, Inc.

Ambler, S. W., & Lines, M. (2017). An Executive's Guide to Disciplined Agile. Calgary: Disciplined Agile Consortium.

Beck, K., & Andres, C. (2005). Extreme Programming Explained: Embrace Change. Upper Saddle River: Pearson Education, Inc.

Boehm, B., Lane, J. A., & Koolmanojwong, S. T. (2014). The Incremental Commitment Spiral Model: Principles and Practices for Successful Systems and Software. Upper Saddle River: Pearson Education, Inc.

Booch, G., Maksimchuck, R. A., Engle, M. W., Yoiung, B. J., Conallen, J., & Houston, K. A. (2007). Object-oriented Analysis and Design with Applications, Third Edition. Boston: Pearson Education, Inc.

Boyde, J. (2014). SDLC Project Management: Getting your Systems Development Life Cycle project successfully across the line using PMBOK... adaptively, Second Edition. San Bernardino: Self Published.

Cockburn, A. (2007). Agile Software Development: The Cooperative Game, Second Edition. Boston: Pearson Education, Inc.

Cockburn, A., & Highsmith, J. (2005). Crystal Clear: A human-powered methodology for small teams. Upper Saddle River: Pearson Education, Inc.

Cohn, M. (2004). User Stories Applied: For Agile Software Development. Boston: Addison-Wesley.

Crookshanks, E. (2017). Just Enough Requirements and SDLC: Requirements documentation, Waterfall and Agile. Virginia Beach: CreateSpace Publishing, Word-2-Kindle.

Font, V. M. (2012). The Ultimate Guide to the SDLC. Raleigh: Font Life Publication, LLC.

Ford, N., Parsons, R., & Kua, P. (2017). Building Evolutionary Architectures. Sebastopol: O'Reily Media, Inc.

Gilb, T. (1977). Software Metrics. Cambridge: Winthrop Publishers, Inc.

Goldratt, E. M., & Cox, J. (2014). The Goal: A process of ongoing improvement, Fourth Revised Edition. Great Barrington: The North River Press Publishing Corporation.

Humphrey, W. S. (2000). Introduction to the Team Software Process. Reading: Computer and Engineering Publishing group, Addison-Wesley Longman, Inc.

Humphrey, W. S. (2005). PSP: A self-improvement process for software engineers. Upper Saddle River: Pearson Education, Inc.

International Institute of Business Analysis (IIBA). (2015). BABOK: A guide to the Business Analysis Body of Knowledge. Toronto: International Institute of Business Analysis (IIBA).

Jacobson, I., Christenson, M., Jonsson, P., & Overgaard, G. (1992). Object-Oriented Software Engineering: A use case driven approach. Harlow, Essex, England: Addison-Wesley Longman Limited.

Kennaley, M. (2010). SDLC 3.0: Beyond a tacit understanding of Agile. Charleston: Fourth Medium Consulting, Inc.

Kerr, J., & Hunter, R. (1994). Inside RAD: How to build fully functional computer systems in 90 days or less. New York: McGraw-Hill, Inc.

Kessler, C., & Sweitzer, J. (2008). Outside-In Software Development: A practical approach to building successful stakeholder-based products. Boston: IBM Press, Pearson plc.

Kim, G., Behr, K., & Spafford, G. (2018). The Phoenix Project: A novel about IT, DevOps, and helping your business win, 5th Anniversary Limited Edition. Portland: IT Revolution.

Kim, G., Humble, J., Debois, P., & Willis, J. (2016). The DevOps Handbook: How to create world-class agility, reliability, & security in technology organizations.

Portland: IT Revolution Press, LLC.

Knaster, R. R., & Leffingwell, D. (2019). SAFe Distilled: Applying the Scaled Agile Framework for Lean Enterprises. Boston: Addison-Wesley.

Ladas, C. (2008). Scrumban: Essays on Kanban Systems for Lean Software Development. Seattle: Modus Cooperandi Press.

Larman, C., & Vodde, B. (2017). Large-Scale Scrum: Moew with less. Boston: Addison-Wesley.

Leffinggwell, D., Knaster, R., Oren, I., & Jemilo, D. (2018). SAFe Reference Guide: Scaled Agile Framework for Lean Enterprises. Boulder: Scaled Agile, Inc.

Leffingwell, D. (2007). Scaling Software Agility: Best practices for Large Enterprises. Boston: Pearson Education, Inc.

Leyton, R. (2015). The Agile Business Analyst: Moving from Waterfall to Agile. San Bernardino: Self Published.

Lines, M., & Ambler, S. W. (2015). Introduction to Disciplined Agile Delivery. Digital Agile Consortium: Calgary.

Martin, J. (1991). Rapid Application Development. New York: Macmillan Publishing Company, Collier Macmillan Canada, Inc.

Poppendieck, M., & Poppendieck, T. (2003). Lean Software Development: An Agile Toolkit. Upper Saddle River: Addison-Wesley.

Poppendieck, M., & Poppendieck, T. (2007). Implementing Lean Software Development: From concept to cash. Boston: Pearson Education, Inc.

Project Management Institute (PMI). (2015). Business Analysis for Practitioners, A Practice Guide. Newtown Square: Project Management Institute, Inc.

Project Management Institute (PMI). (2017). A Guide to the Project Management Body of Knowledge: PMBOK Guide, Sixth Edition. Newtown Square: Project Management Institute.

Project Management Institute, Inc. (PMI) . (2017). Agile Practice Guide, Global Standard. Newtown Square: Project Management Institute, Inc.

Rock-Evans, R. (1992). Data Modeling & Process Modeling. Oxford: Butterworth-Heinemann, Ltd.

Sharma, S. (2017). The DevOps Adoption Playbook. Indianapolis: John Wiley & Sons, Inc.

Shewhart, W. A. (1986). Statistical Method from the viewpoint of Quality Control. New York: Dover Publications, Inc.

Sutherland, J., & Sutherland, J. (2014). SCRUM: The art of doing twice the work in half the time. New York: Crown Publishing Group, a division of Penguin Random House, LLC.

Weaver, P. L., Lambrou, N., & Walkley, M. (1998). Practical SSADM 4: A complete tutorial guide, second edition. Harlow, Essex, England: Pearson Education Limited.

Weaver, P., Lambrou, N., & Walkley, M. (2002). Practical Business Systems Development Using SSADM: A complete technical guide, Third Edition. Harlow, Essex, England: Pearson Education, Limited.

Wilson, L. B., & Clark, R. G. (2001). Comparative Programming Languages, Third Edition. Essex: Pearson Education Limited.

Womack, J. P., & Jones, D. T. (2003). Lean Thinking: Banish waste and create wealth in your corporation. New York: Free Press, a Division of Simon & Schuster, Inc.

Yakyma, A. (2016). The Rollout: A novel about leadership and building a Lean-agile Enterprise with SAFe. Unknown: Self Published.

Bibliography

1 2019 CollabNet | VersionOne. 12th & 13th Annual State of Agile Reports. Retrieved from https://explore.versionone.com/state-of-Agile/versionone-12th-annual-state-of-Agile-report

2 2019 CollabNet | VersionOne. 12th & 13th Annual State of Agile Reports. Retrieved from https://explore.versionone.com/state-of-Agile/versionone-12th-annual-state-of-Agile-report

3 ©2001 Jim Highsmith. History: The Agile Manifesto. http://agilemanifesto.org/history.html. Accessed 10 April 2019.

4 Kent Beck, et al. © 2001. Manifesto for Agile Software Development. http://agilemanifesto.org/. Accessed 10 April 2019.

5 Kent Beck, et al. © 2001. Principles behind the Agile Manifesto. http://agilemanifesto.org/principles.html. Accessed 10 April 2019.

6 Kent Beck, et al., Manifesto for Agile Software Development. http://agilemanifesto.org

7 Kent Beck, et al., Manifesto for Agile Software Development. http://agilemanifesto.org

8 ©2020. SDLC Forms Documentation Consultants www.sdlcforms.com Accessed 20 January 2020

9 Kent Beck, et al., Manifesto for Agile Software Development. http://agilemanifesto.org

10 Kent Beck, et al., Manifesto for Agile Software Development. http://agilemanifesto.org

11 Benington, Herbert D. (October-December 1983). "Production of Large Computer Programs." IEEE Annals of the History of Computing 5(4), pp. 350-361.

12 Winston, Royce. (1970). "Managing the Development of Large Software Systems." Proceedings of IEEE WESCON.

13 Bell T. E., Thayer T. A. 1976. SOFTWARE REQUIREMENTS: ARE THEY REALLY A PROBLEM. TRW Defense and Space Systems Group Redondo Beach, California. Retrieved from https://pdfs.semanticscholar.org/a50d/8f564d063e526e94114875220440f64d8666.pdf. Accessed 20 January 2020.

14 Military standard DOD-STD-2167. Available from EverySpec LLC, http://everyspec.com/DoD/DoD-STD/DOD-STD-2167A_8470/. Accessed January 26, 2020.

15 MIL-STD-498. Available from EverySpec LLC, http://everyspec.com/MIL-STD/MIL-STD-0300-0499/MIL-STD-498_25500/ Accessed January 26, 2020.

16 Cockburn, Alistair (2008). Using both incremental and iterative development. CrossTalk, The Journal of Defense Software Engineering 21(5), pp. 27–30. Retrieved from http://www.se.rit.edu/~swen-256/resources/UsingBothIncrementalandIterativeDevelopment-AlistairCockburn.pdf Accessed January 26, 2020.

17 Winston, Royce. (1970). "Managing the Development of Large Software Systems." Proceedings of IEEE WESCON.

18 Moen, R. (2009) Foundation and History of the PDSA Cycle. https://deming.org/uploads/paper/PDSA_History_Ron_Moen.pdf Accessed Hanuary 27, 2020.

19 Gilb, T. (2005) Competitive Engineering: A Handbook For Systems Engineering, Requirements Engineering, and Software Engineering Using Planguage 1st Edition. Elsevier Butterworth-Heinemann Publishing. Burlington, MA.

20 R. Zultner, "The Deming Approach to Quality Software Engineering," Quality Progress, vol. 21, no. 11, 1988

21 Larman, Craig & Basili, Victor R. "Iterative and Incremental Development: A Brief History." IEE Computer Society. Issue No. 06 - June (2003 vol. 36), ISSN: 0018-9162, pp: 47-56. http://doi.ieeecomputersociety.org/10.1109/MC.2003.1204375

22 Martin, J. (1991) Rapid Application Development. High Speed, High Quality, Low Cost. MacMillan Publishing Company, New York, NY

23 ©2001 Jim Highsmith. History: The Agile Manifesto. http://agilemanifesto.org/history.html. Accessed January 26, 2020.

24 Kent Beck, et al. © 2001. Manifesto for Agile Software Development. http://agilemanifesto.org/. Accessed January 26, 2020.cfm?LeanPersonId=1

25 SCRUM Guide. ©2018 Ken Schwaber and Jeff Sutherland. ScrumGuides.org Offered for license under the Attribution Share-Alike license of Creative Commons, accessible at https://www.scrumguides.org/. Accessed January 26, 2020.

26 Modern Agile. http://modernAgile.org/ Content available for use under the Creative Commons Attribution-ShareAlike 4.0 International License. http://creativecommons.org/licenses/by-sa/4.0/.

27 Broadbent, D. (1958). Perception and Communication. London: Pergamon Press.

28 The publications of Wolfgang Mewes are marketed through the FAZ Verlag, publishing house of the German newspaper Frankfurter Allgemeine Zeitung.

29 Booch, G., Maksimchuk, R., Engle, M., Young, B., Conallen, J., Houston, K. (1993). Object-oriented analysis and design with Applications (3rd ed.). Upper Saddle River, NJ. Addison Wesley.

30 Beck, K., Andres, C. (2004). Extreme programming explained: Embrace change, US Edition. Boston, MA: Addison Wesley.

31 Cunningham, W. (2009). Integration Hell. Retrieved from http://wiki.c2.com/?IntegrationHell

32 Rajkumar, S. M. (2017). 100+ Types of software testing—The ultimate list | Software testing material. Retrieved from https://www.softwaretestingmaterial.com/

33 Poppendieck, M., & Poppendieck, T. (2003). Lean Software Development: An Agile Toolkit. Addison-Wesley Professional.

34 © Copyright 2000-2019 Lean Enterprise Institute, Inc. James P. Womack biography. Retrieved from: https://www.lean.org/WhoWeAre/LeanPerson.cfm?LeanPersonId=1

35 © Copyright 2000-2019 Lean Enterprise Institute, Inc. Senior Advisors. Retrieved from: https://www.lean.org/WhoWeAre/People.cfm.

36 © Copyright 2000-2019 Lean Enterprise Institute, Inc. All rights reserved. What is Lean? Retrieved from: https://www.lean.org/WhatsLean/

37 © Copyright 2000-2019 Lean Enterprise Institute, Inc. All rights reserved. What is Lean? Retrieved from: https://www.lean.org/WhatsLean/D36

38 Womack, J. P. & Jones, D. T. (2003, revised 2010). Lean thinking: Banish waste and create wealth in your corporation. New York, NY. Simon and Schuster.

39 Toyota Motor Corporation (2019). Toyota production system. Retrieved from corporate website at: https://www.toyota-global.com/company/vision_philosophy/toyota_production_system/

40 Toyota Motor Corporation (2019). Toyota production system. Retrieved from corporate website at: https://www.toyota-global.com/company/vision_philosophy/toyota_production_system/

41 Ibid. Toyota production system.

42 Ibid. Toyota production system.

43 Toyota (GB) PLC. (2013). The Official Blog of Toyota GB. Toyota production system glossary. Retrieved from https://blog.toyota.co.uk/toyota-production-system-glossary

44 Toyota (GB) PLC. (2013). The Official Blog of Toyota GB. Kaizen–Toyota production system guide. Retrieved from https://blog.toyota.co.uk/kaizen-toyota-production-system

45 Lean Global Network. (2015). Planet Lean – The Lean Global Network Journal. PL interviews Jeff Sutherland, one of the inventors of scrum. Retrieved from https://planet-lean.com/jeff-sutherland-scrum-interview/

46 Ladas, C. (2019). [Blog post]. Retrieved from http://leansoftwareengineering.com/ksse/scrum-ban/

47 Ladas, C. (2008). Scrumban: Essays on Kanban Systems for Lean Software Development. Modus Cooperandi Press.

48 Ladas, C. (2008). Scrumban: Essays on Kanban Systems for Lean Software Development. Modus Cooperandi Press.

49 Ladas, C. (2008). Scrumban: Essays on Kanban Systems for Lean Software Development. Modus Cooperandi Press.

50 De Luca, J. Coad, P., & Lefebvre, P. (1999). Java modeling in color with UML. Upper Saddle River, NJ. Prentice Hall PTR

51 Codd, E. F. (1970). A relational model of data for large shared data banks. Communications of the ACM 13(6) p. 377. doi:10.1145/362384.362685.

52 Chen, P. (1976). The entity-relationship model—Toward a unified view of data. ACM Transactions on Database Systems 1(1), pp. 9–36. doi:10.1145/320434.320440.

53 [48] Mayer, R. J. (1995). Information Integration for Concurrent Engineering (IICE) Compendium of methods report. Wright-Patterson Air Force Base, Ohio 45433-7604. Downloads available at KBSI, Inc. http://www.idef.com/downloads/ Copyright © 2019 Knowledge Based Systems, Inc.1408 University Dr. · College Station · Texas 77840 United States. Accessed January 2019.

54 Mayer, R.J., Crump, J.W., Fernandes, F., Keen, A., Painter, M.K. (1995) INFORMATION INTEGRATION FOR CONCURRENT ENGINEERING (IICE) COMPENDIUM OF METHODS REPORT. Defense Technical Information Center. https://apps.dtic.mil/dtic/tr/fulltext/u2/a531128.pdf Accessed January 26, 2020.

55 Ambler, Scott. "The Agile Unified Process (AUP)". Ambysoft. Copyright 2005-2015. http://www.ambysoft.com/unifiedprocess/AgileUP.html. Accessed December 2018.

56 Kurtz, C, F., & Snowden, D, J. (2003). The new dynamics of strategy: Sense-making in a complex and complicated world. IBM Systems Journal 42(3): 462–483. doi:10.1147/sj.423.0462

57 Snowden, D. (2008). Cognitive Edge. About SenseMaker® Retrieved from: https://cognitive-edge.com/sensemaker/#sensemaker-about

58 Snowden, D. (1999). "Liberating knowledge." Introductory Chapter CBI Business Guide. London: Caspian Publishing.

59 Snowden, D. (October 2017). Cognitive edge/Liminal Cynefin: the final cut? Retrieved from http://cognitive-edge.com/blog/liminal-cynefin-the-final-cut/

60 Snowden, D. (June 2008). Cognitive Edge/What is Sense-making? Retrieved from http://cognitive-edge.com/blog/what-is-sense-making/.

61 Snowden, D. (July 2002). Complex Acts of Knowing—Paradox and Descriptive Self Awareness. Journal of Knowledge Management, Special Issue. Retrieved from http://cognitive-edge.com/articles/complex-acts-of-knowing-paradox-and-descriptive-self-awareness/

62 Girard, J. P. & Girard, J. L. (2015). "Defining knowledge management: Toward an applied compendium" (PDF). Online Journal of Applied Knowledge Management. 3(1), p. 14.

63 Lines, M., & Ambler, S. (2018). Introduction to disciplined Agile delivery 2nd Edition: A small team's journey from Scrum to disciplined DevOps. North Charleston, SC. Disciplined Agile Consortium.

64 Ambler, S. & Lines, M. (2017). An executive guide to disciplined Agile: Winning the race to business agility. Disciplined Agile Consortium. North Charleston, SC. CreateSpace Independent Publishing Platform

65 Ambler, S. & Lines, M. (2012). Disciplined Agile delivery: A practitioner's guide to Agile software delivery in the enterprise. Upper Saddle River, NJ. IBM press. Pearson PLC.

66 Ambler, S. & Lines, M. (2019). Choose your WoW!: A disciplined Agile delivery. Foundations handbook for optimizing your way of working (WoW) North Charleston, SC. CreateSpace Independent Publishing Platform.

67 Lines, M., & Ambler, S. (2019) Welcome to the Disciplined Agile Membership Site. 2014-2018 Disciplined Agile Consortium. Retrieved from https://www.disciplinedagileconsortium.org/ Accessed January 27, 2020.

68 Ambler, S. (2019). Agile Modeling (AM) Home Page. Effective Practices for Modeling and Documentation. Retrieved from http://agilemodeling.com/ Accessed January 26, 2020.

69 Ambler, S. (2019). Agile Data Home Page. Retrieved from http://agiledata.org/ Accessed January 26, 2020.

70 Ambler, S. (2019) Enterprise Unified Process (EUP): Strategies for Enterprise Agile. http://enterpriseunifiedprocess.com/ Accessed January 26, 2020.

71 Lines, M., & Ambler, S. (2019) Choose Your WoW! A Disciplined Agile Delivery Handbook for Optimizing Your Way of Working (WoW). Calgary, AB. Independently Published by Disciplined Agile, Inc

72 Corporate Author. (2019) Agile Business Consortium Limited. What is DSDM? Retrieved from https://www.agilebusiness.org/what-is-dsdm

73 Beck, K., Andres, C. © November 2004. Extreme Programming Explained: Embrace Change, 2nd Edition (The XP Series). 2nd Edition.

74 15 Software Development Influencers You Should Follow in 2018. by Ekaterina Novoseltseva · Dec. 27, 17 · DZone/ Agile Zone · Opinion. https://dzone.com/articles/15-software-development-influencers. Accessed December 2018.

75 Kessler, C. & Sweitzer, J. (September 2008). Outside-in software development: A practical approach to building successful stakeholder-based products, Upper Saddle River, NJ. IBM Press. Pearson PLC.

76 Franz, A. (2015). Outside-in vs. inside-out thinking. Retrieved from http://customerthink.com/outside-in-vs-inside-out-thinking/

77 CMMI Institute LLC (2019). An ISACA Enterprise: About CMMI. Retrieved from https://cmmiinstitute.com/company

78 Callison, R. & MacDonald, M. (October 2009). Special report CMU/SEI-2009-SR-025. Software Engineering Institute. Retrieved from https://resources.sei.cmu.edu/asset_files/SpecialReport/2009_003_001_15038.pdf

79 CMU/SEI Digital Library. Retrieved from https://resources.sei.cmu.edu/library/results.cfm?as_q=inmeta:gsataxonomyoutput~TSP

80 Humphrey, W., (2005) PSPsm – A self-improvement Process for Software engineers. Upper Saddle, NJ. Addison-Wesley/ Pearson Education.

81 Humphrey, W., (2005) TSPsm – Introduction to the Team Software Process. Upper Saddle, NJ. Addison-Wesley/ Pearson Education.

82 Leffingwell, D. (2007) Scaling Software Agility: Best Practices for Large Enterprises. 1st Edition. Upper Saddle River, NJ.

83 Scaled Agile, Inc. Provider of SAFe. (2019) SAFe® for Lean Enterprises. Retrieved from https://www.scaledagileframework.com/.

84 Takeuchi, H. & Nonaka, I. (January 1, 1986). "The new new product development game". Harvard Business Review. Moving the Scrum Downfield

85 Sutherland, J. V. & Schwaber, K. (1995). Business object design and implementation: OOPSLA '95 workshop proceedings, p. 118. The University of Michigan.

86 Schwaber, K. & Sutherland, J. (2018). Scrum Guide. Retrieved from https://www.scrumguides.org/ Accessed January 26, 2020.

87 Boehm B. (August 1986). A spiral model of software development and enhancement. ACM SIGSOFT software engineering notes. ACM 11(4), pp. 14-24.

88 Boehm B. (May 1988). A spiral model of software development and enhancement. IEEE Computer. IEEE 21(5), pp. 61-72.

89 Corporate Author. (1993) IABG Information Technology. V-Model Life cycle Process Model. Archived by Internet WayBack Machine from the original on March 3, 2016. Retrieved from http://www.v-modell.iabg.de/kurzb/vm/k_vm_e.doc

90 2019 CollabNet | VersionOne. 12th & 13th Annual State of Agile Reports. Retrieved from https://explore.versionone.com/state-of-Agile/versionone-12th-annual-state-of-Agile-report

91 The Object Management Group (OMG) (2019) Home Page. Retrieved from https://www.omg.org/ Accessed January 26, 2020

92 CollabNet | VersionOne. 12th & 13th Annual State of Agile Reports. Retrieved from https://explore.versionone.com/state-of-Agile/versionone-12th-annual-state-of-Agile-report Accessed January 26, 2020.

93 12207-2017 - ISO/IEC/IEEE International Standard - Systems and software engineering -- Software life cycle processes. November 2017. Standards Catalogue. Retrieved from https://ieeexplore.ieee.org/document/8100771.

94 IEEE 15288-2015 - ISO/IEC/IEEE International Standard - Systems and software engineering -- System life cycle processes Standards Catalog. Retrieved from https://standards.ieee.org/standard/15288-2015.html

95 Sarbanes Oxley Act 2002, Section 404. Retrieved from http://www.soxlaw.com/s404.ht

96 Health IT Legislation. Retrieved from https://www.healthit.gov/topic/laws-regulation-and-policy/health-it-legislation. Accessed on January 5, 201

97 TechTarget. (2000). Introduction to COBIT for SOX compliance. Retrieved from http://searchsecurity.techtarget.com/feature/Introduction-to-COBIT-for-SOX-compliance

98 Wikipedia. (2018) List of computer standards. Text is available under the Creative Commons Attribution-ShareAlike License. Retrieved from https://en.wikipedia.org/wiki/List_of_computer_standards. Accessed on January 5, 2019

99 WeboPedia. (2019). An eWEEK property. https://www.webopedia.com/Networks/Networking_Standards. Accessed on January 5, 2019

100 IT Governance USA Inc. (2019). IT Standards. Retrieved from https://www.itgovernanceusa.com/standards Accessed on January 5, 2019.

101 Documentation Consultants. Home Page. www.sdlcforms.com Accessed January 26, 2020.

102 VersionOne CollabNet (April 9, 2018). 13th Annual State of Agile Report/ Percentage of Teams Using Agile. Page 7. Retrieved from https://explore.versionone.com/state-of-Agile/versionone-12th-annual-state-of-Agile-report Accessed January 26, 2020.

103 Benington, H. D. (1983). Production of large computer programs, Issue No. 4, vol. 5. ISSN: 1058-6180, pp: 350-361. doi: http://doi.ieeecomputersociety.org/10.1109/MAHC.1983.10102

104 Larman, C. & Basili, V. R. (2003). Iterative and Incremental Development: A Brief History. IEEE Computer 36 (3) pp. 47–56. doi:10.1109/MC.2003.1204375

105 Department of Defense. (1988). Defense Systems Software Development: Standard 2167A (DOD STD-2167A). Retrieved from http://military.wikia.com/wiki/DOD-STD-2167A

106 Wikipedia. (2015) MIL STD 498. Retrieved from https://en.wikipedia.org/wiki/MIL-STD-498

107 Internet Archive. Wayback Machine. Data Item Description. http://www.fisica.uniud.it/~cabras/swe/MIL-STD/498/SDP-DID.PDF Accessed January 26, 2020.

108 IBID

109 Project Management Institute (2017) A Guide to the A Guide to the Project Management Body of Knowledge (PMBOK® Guide)–Sixth Edition. Pages 23-25. Newtown Square, PA Project Management Institute.

110 Shafer, Tom. (2017). The 42 V's of Big Data and Data Science. Retrieved from https://www.elderresearch.com/blog/42-v-of-big-data

111 taxonomy. (n.d.). In Collins English Dictionary (10th ed.). Retrieved from http://www.dictionary.com/browse/taxonomy

112 IEEE Website, © Copyright 2018 IEEE – All rights reserved. https://www.ieee.org/ Accessed April 20, 2018.

113 Object Management Group. OMG Specifications Catalog. Retrieved from https://www.omg.org/spec/ Accessed January 26, 2020.

114 The Open Group. (2019). Standards. Retrieved from https://publications.opengroup.org/standards?p=1. Accessed on January 5, 2019

Acknowledgments

While I enjoy working in the field of information technology, we can never forget that "people work with people." One of my early mentors, Jim O'Malley, told me that when I was a young consulting engineer supporting his region in a startup software company. While we continue to build our digital world, we can't lose sight of this most basic human element.

Throughout my career, I've had the pleasure of working with many extremely capable information technology, domain and business specialists. Virtually all of the experience and skills I've gained resulted from endless hours of collaboration and problem solving with these fine folks, and I've tried to stay in touch. You just never know when you might need to hear their thoughts and opinions or ask them to review the draft of your latest book.

Truly, this book would not have been possible without the help of a host of people who were kind enough to read through the drafts and provide their thoughts and opinions, such as Rebecca Ames, Jessica Anderson, Patricia F. Diaz, PhD, Martha Evert, Enrique Gomez, Ed Hicks, Rich Homes, Kathie Jacobsen, Michael Krause, David Rice, Ed Ridley, PhD, and Mark Zalubas. I may have missed some folks, and I apologize if I have.

In addition, I would like to thank the many industry experts whom I reached out to for their input, assistance with cite permissions, and professional advice on certain subjects, to include Paul Derstine of Elder.com, Geoff Terrell of the CMMI Institute, Jodi Lebow of Collabnet, D. J. Snowden of Cognitive Edge, Scott Ambler of Disciplined Agile, Sharon Corbett of the Eclipse Foundation, M.E. Brennan of IEEE, Paul Stapleton of IIBA, Jerry Weinberg (deceased), Bas Vodde and Craig Larman of LeSS, Martin Fowler of ThoughtWorks and martinfowler.com, John Sterman - Director, MIT System Dynamics Group – Sloan School of Management, Mike Cohn of Mountain Goat Software, Catherine Viega of the Project Management Institute (PMI), Adam Mattis and Inbar Oren of Scaled Agile, Robert DeLaet of SDLCforms, Vijay Shinde and Kamila of Software Testing Help, Veronika Kiesswetter of AXELOS Global Best Practice, Margaret Rouse of TechTarget, Joyce Brown and Meryl Schlachterman of The Open Group, Paul Jansen of TIOBE Software BV, James Grenning of Wingman Software and Pramod Malhotra of Salient CRGT. Again, my sincerest apologies if I've missed anyone. In all cases, I have endeavored to be accurate in my representations of their comments or requests that I've incorporated within this book.

About the Author

Cecil "Gary" Rupp brings more than twenty-five years of executive-level experience in information technology (IT), spanning consulting, marketing, professional services, program and project management, and sales. Mr. Rupp has directly managed more than thirty enterprise-class IT programs and projects, with the last ten years focused almost exclusively on managing large enterprise Health IT Programs. His professional experience includes supporting large software and systems development projects and promoting and delivering IT solutions involving the use of computer-aided systems engineering (CASE), middleware, portal, and software development tools. He is the author of the *Building our Digital World* series of books on software and systems development practices.

Professional Certifications
Scaled-Agile, Inc. Certified SAFe® Program Consultant ID No. 24488227-0897.
PMI certified Project Management Professional (PMP) – PMP Number: 243644.
Scrum Alliance Certified Scrum Master (CSM) – ID No. 650996.
Scrum Alliance Certified Scrum Product Owner (CSPO) ID No. 650996.
Certified Large-Scale Scrum (LeSS) Practitioner.

Made in United States
North Haven, CT
05 November 2026

21764038R00137